Make it easy...

Maths

Age 8–9

Paul Broadbent

Numbers to *1000*

The numbers between 100 and 999 all have **three digits**.

376 → 300 + 70 + 6

(3 = hundreds, 7 = tens, 6 = ones)

When you add or subtract 1, 10 or 100, the digits change.

376 + 1 = 37**7**   376 + 10 = 3**8**6   376 + 100 = **4**76

I Continue these number chains.

a	757 →	+ 1	→ 758 →	+ 1	→ 759 →	+ 1	→ 760
b	628 →	– 10	→ 618 →	– 10	→ 608 →	– 10	→ 598
c	496 →	+ 10	→ 506 →	+ 10	→ 516 →	+ 10	→ 526
d	641 →	+ 100	→ 741 →	+ 100	→ 841 →	+ 100	→ 941
e	385 →	– 100	→ 285 →	– 100	→ 185 →	– 100	→ 85
f	903 →	– 1	→ 902 →	– 1	→ 901 →	– 1	→ 900

II Complete this number puzzle.

Across

1 Seven hundred and forty-three

5 Nine hundred and twenty

Down

2 Four hundred and nine

3 Three hundred and fifty-one

4 Six hundred and eight

Number sequences

A sequence is usually a list of **numbers in a pattern.**

Look at the difference between each number to spot the rule for the pattern.

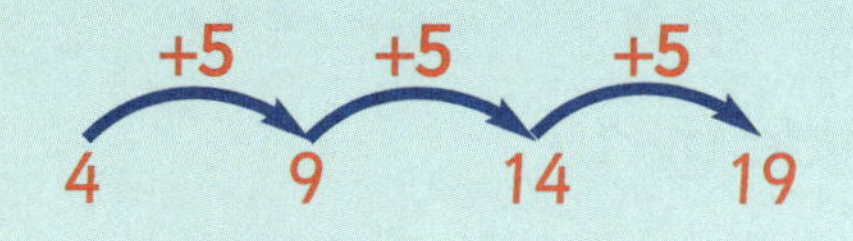

The rule is +5

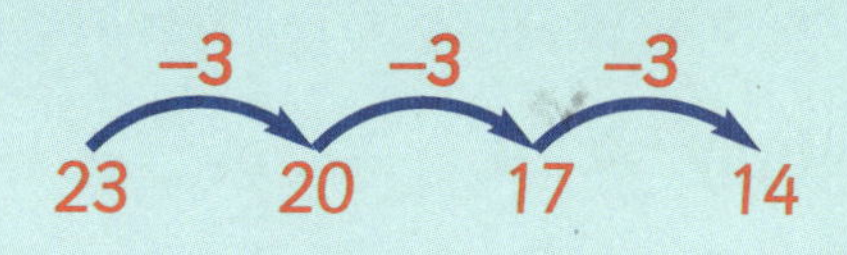

The rule is −3

I Write the missing numbers in these sequences. What is the rule for each of them?

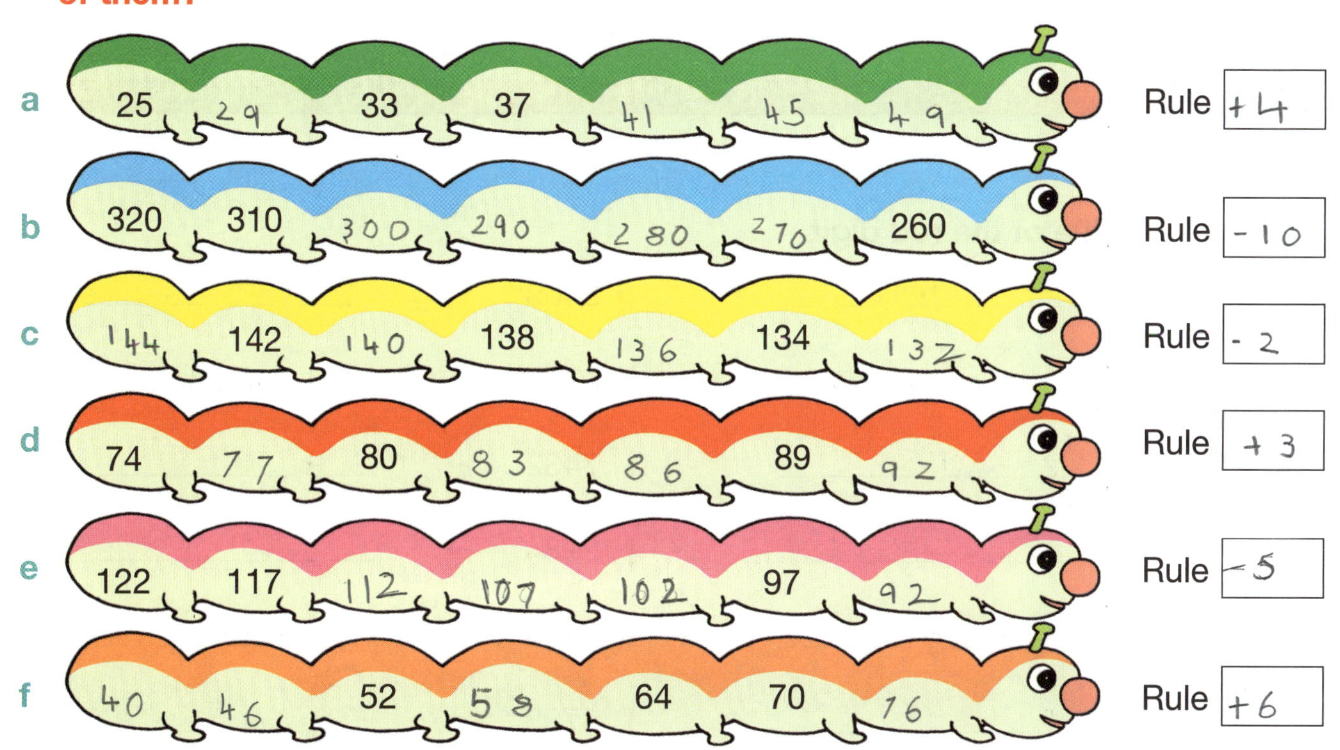

II Negative numbers go back past zero. Write the missing numbers on these number lines.

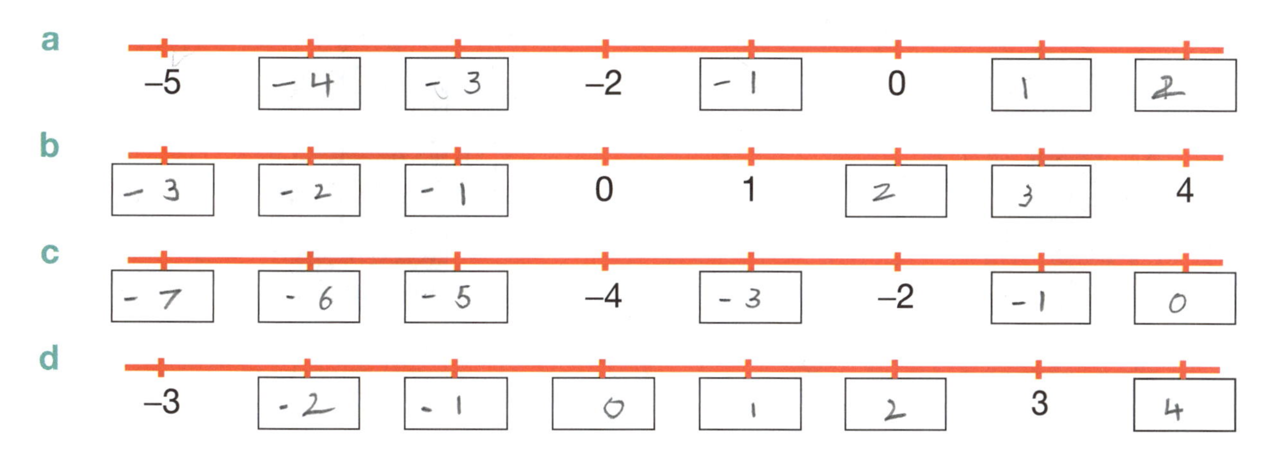

Place value

4-digit numbers are made from **thousands, hundreds, tens** and **ones.**

To make a number ten times bigger or ten times smaller you need to move the digits.

Multiply by 10

- Move the digits **one** place to the left.
- Fill the spaces with zero.

	3	8	2	× 10
3	8	2	0	

Divide by 10

- Move the digits **one** place to the right.

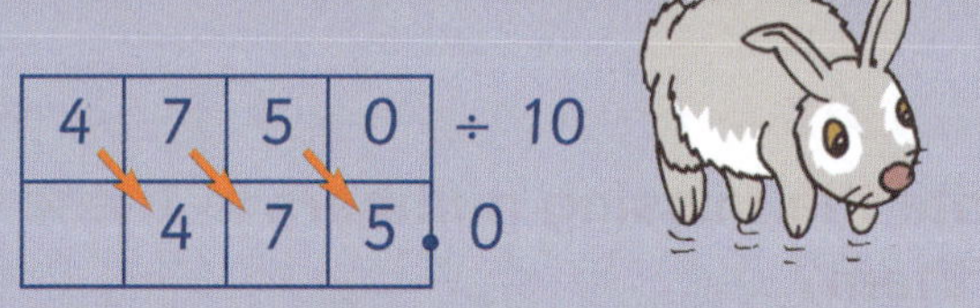

I Write the value of the red digit.

a 3450 → 5 tens

b 6795 → 6 thosunds

c 4008 → 8 units

d 9217 → 2 hunndreds

e 3169 → 6 tens

f 5291 → 5 thosunds

g 9469 → 9 units

h 4778 → 7 humndreds

i 7432 → 3 tens

j 2984 → 4 units

k 8898 → 8 thosunds

l 4793 → 7 hunndreds

II Write the numbers coming out of each machine.

× 10

a 385 → 3850

b 790 → 7900

c 368 → 3680

d 412 → 4120

e 900 → 9000

÷ 10

f 4650 → 465

g 2910 → 291

h 3400 → 340

i 5070 → 507

j 8000 → 800

Addition and subtraction

If you learn the addition and subtraction facts to 20, they can help you to learn other facts. Look at these patterns.

4 + 9 = 13	15 − 8 = 7
40 + 90 = 130	150 − 80 = 70
400 + 900 = 1300	1500 − 800 = 700

I Write the answers to these questions.

a
7 + 5 = 12
70 + 50 = 120
700 + 500 = 1200

b
9 + 6 = 15
90 + 60 = 150
900 + 600 = 1500

c
4 + 11 = 15
40 + 110 = 150
400 + 1100 = 1500

d
13 − 6 = 7
130 − 60 = 70
1300 − 600 = 700

e
15 − 7 = 8
150 − 70 = 80
1500 − 700 = 800

f
18 − 9 = 9
180 − 90 = 90
1800 − 900 = 900

g 180 − 60 = 120

h 800 + 500 = 130

i 1700 − 400 = 1300

j 1200 + 600 = 1800

k 150 + 90 = 1600

l 130 − 90 = 40

m 800 + 800 = 1600

II Circle touching pairs of numbers that total 100. The pairs can be vertical or horizontal. You should find ten pairs.

34	51	59	41	76	82	38	62
66	75	25	65	24	47	53	77
91	19	72	83	17	45	96	13
24	81	74	56	35	55	48	52

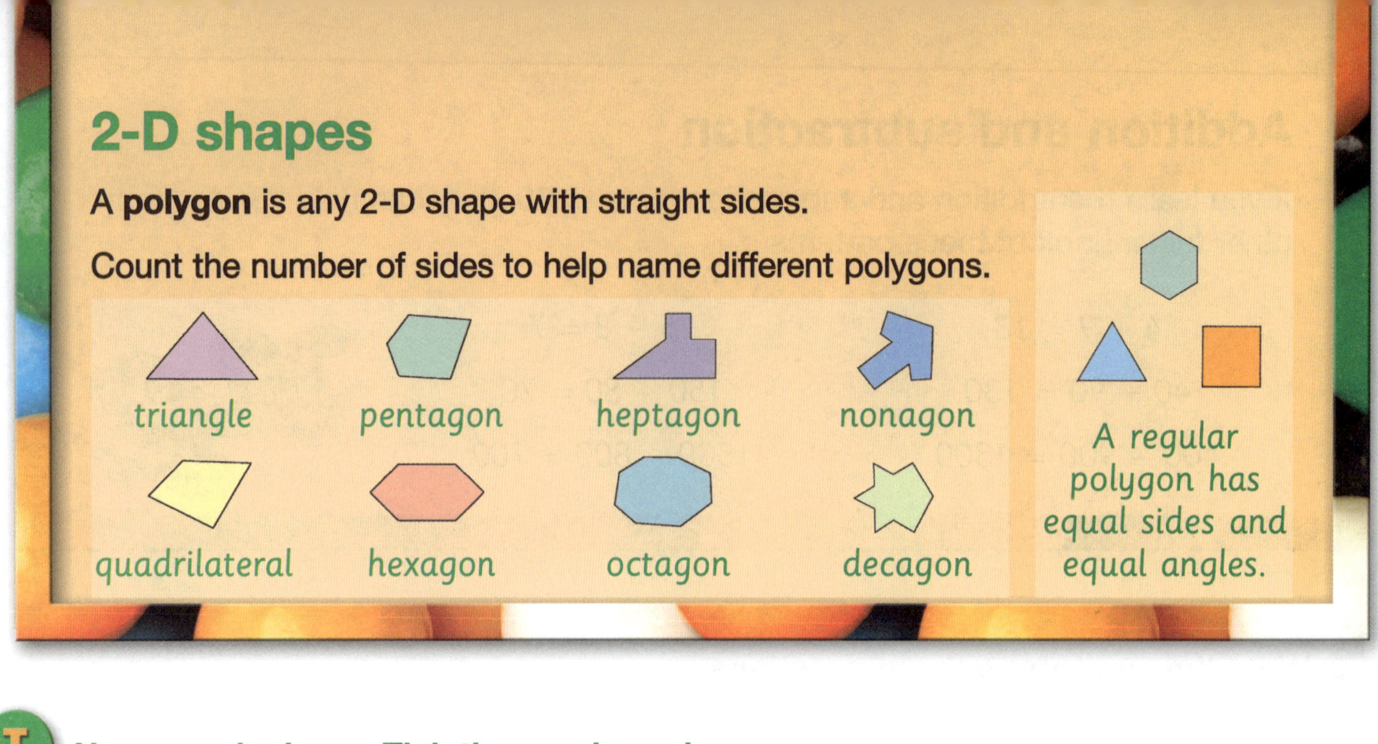

2-D shapes

A **polygon** is any 2-D shape with straight sides.

Count the number of sides to help name different polygons.

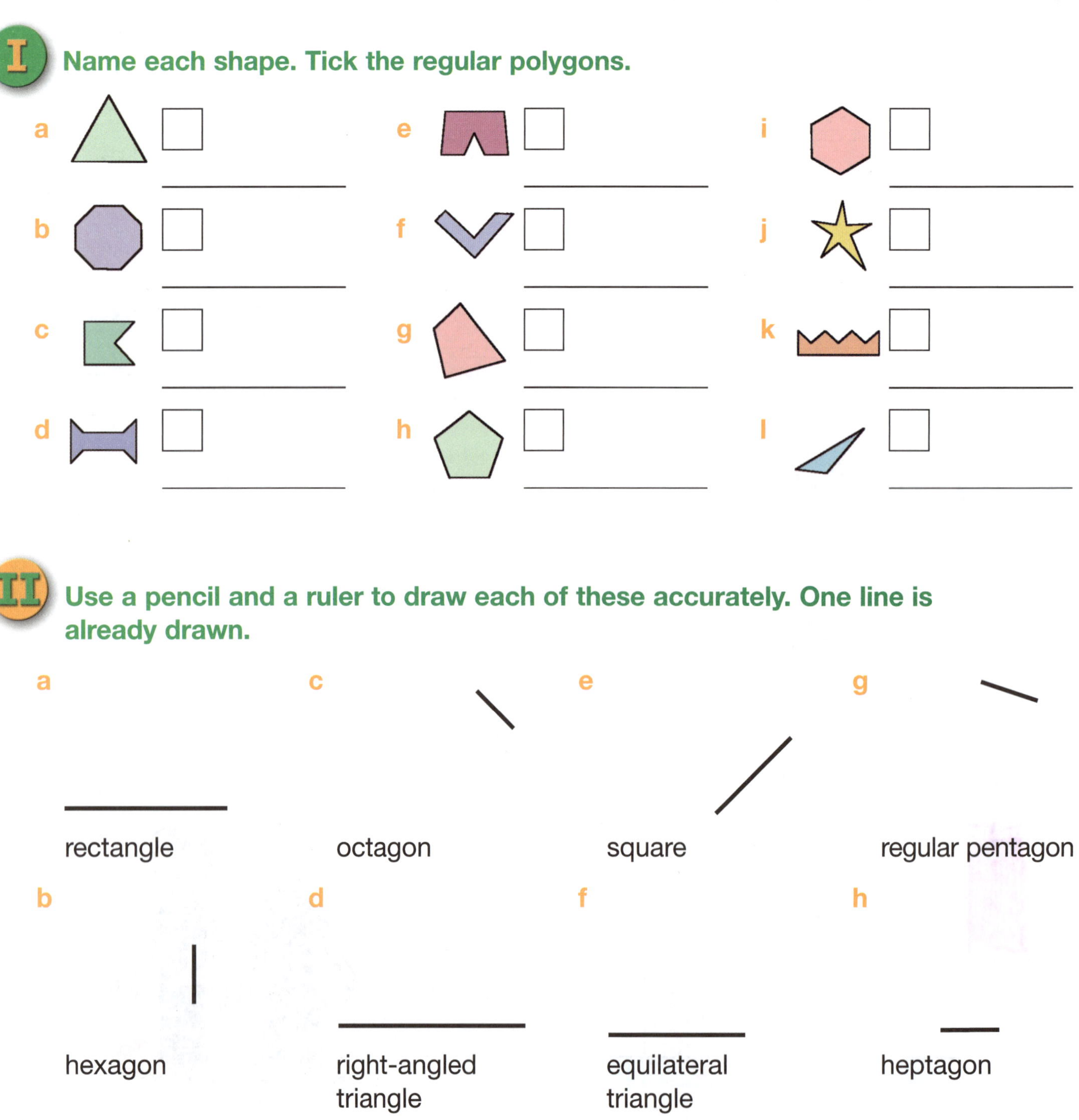

I Name each shape. Tick the regular polygons.

a b c d e f g h i j k l

II Use a pencil and a ruler to draw each of these accurately. One line is already drawn.

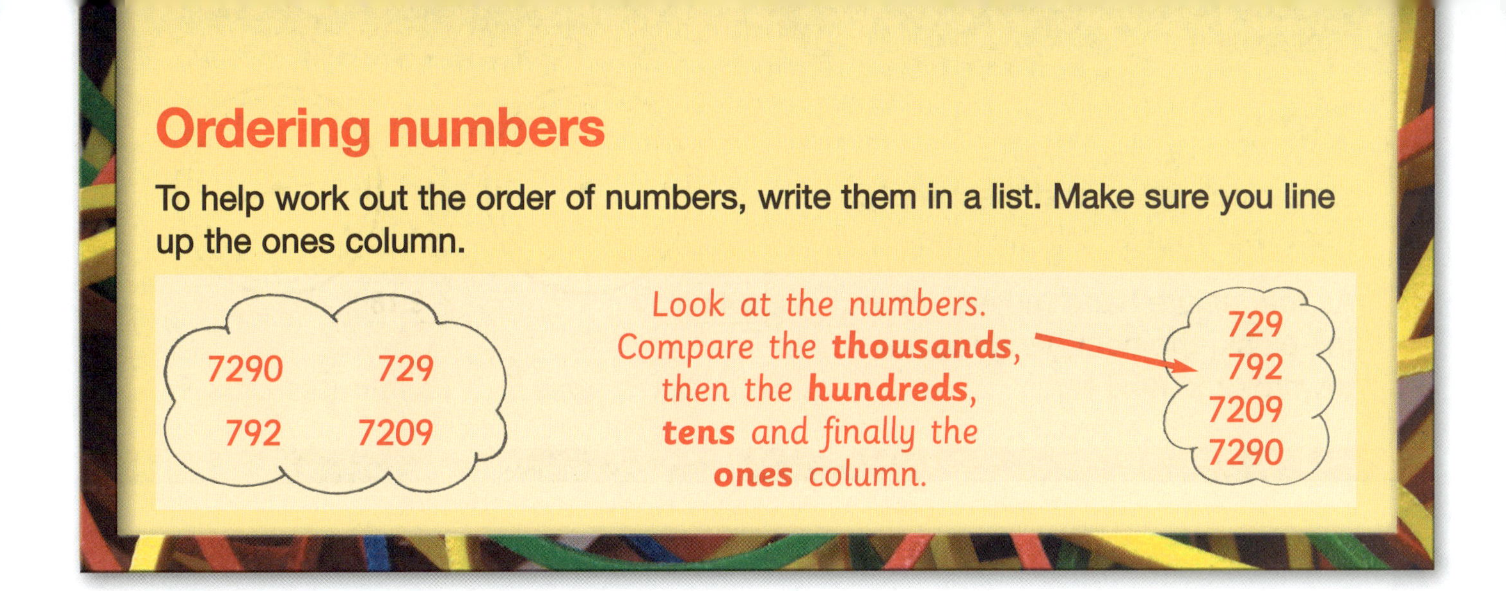

Ordering numbers

To help work out the order of numbers, write them in a list. Make sure you line up the ones column.

I Write these in order, starting with the smallest.

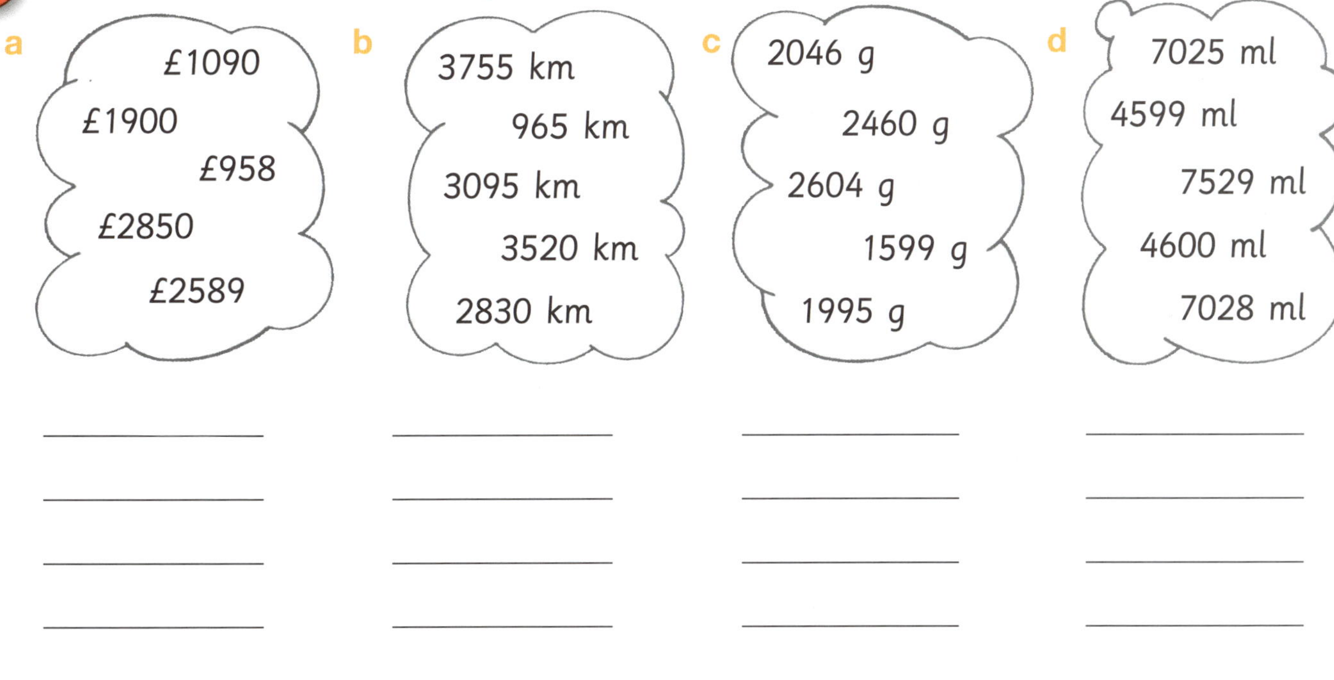

II Use the digits 2 9 3 8.

Make as many different 4-digit numbers as you can. Write them in order, starting with the smallest.

Time

On a clock face, read the **minutes past the hour** to tell the time.

As the minute hand moves around the clock, the hour hand moves towards the next hour.

42 minutes past 5 18 minutes past 9

I Write the times shown on each clock.

a 

12 past 4

b 

11 to 11

c 

26 past 12

d 

25 to 6

Draw the hands on these clocks.

e 

7.56

f 

9.03

g 

3.41

h 

11.18

II Write the number of minutes between each of these times.

a 

25 minutes

c  

50 minutes

b 

80 minutes

d 

75 minutes

Fractions of amounts

The number below the line of a fraction tells you how many parts to **divide** into.

$\frac{1}{4}$ of 8 is the same as 8 ÷ **4** = 2

$\frac{1}{3}$ of 15 is the same as 15 ÷ **3** = 5

I Use the pictures to help you answer these problems.

a

b

c

a	b	c
$\frac{1}{2}$ of 12 = 6	$\frac{1}{2}$ of 20 = 10	$\frac{1}{2}$ of 24 = 12
$\frac{1}{4}$ of 12 = 3	$\frac{1}{4}$ of 20 = 5	$\frac{1}{3}$ of 24 = 6
$\frac{1}{3}$ of 12 = 4	$\frac{1}{5}$ of 20 = 4	$\frac{1}{4}$ of 24 = 3
$\frac{1}{6}$ of 12 = 2	$\frac{1}{10}$ of 20 = 16	$\frac{1}{6}$ of 24 = 4

II Colour these grids to match the fractions. How many squares are left white on each? Make interesting patterns on each grid.

$\frac{1}{2}$ → red

$\frac{1}{4}$ → blue

$\frac{1}{6}$ → yellow

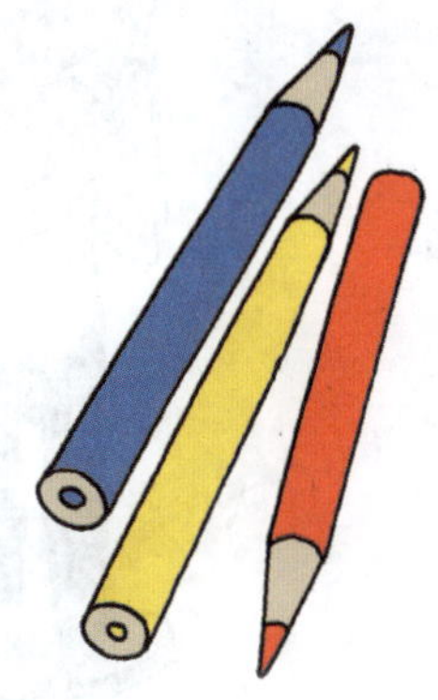

a

3 left white

b

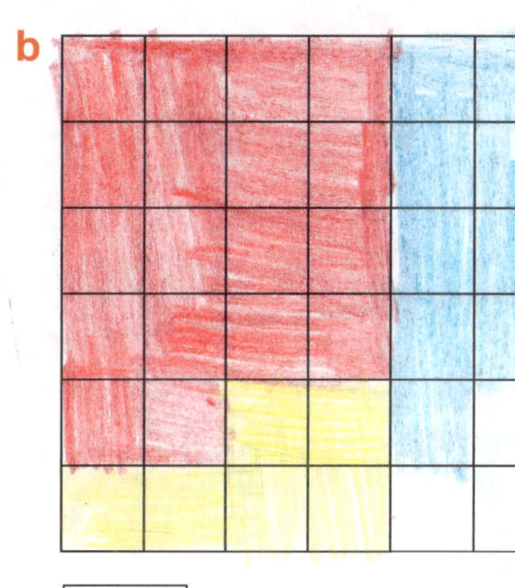

15 left white

Measuring length

Look at these lengths.

10 millimetres (mm)	= 1 centimetre (cm)
100 cm	= 1 metre (m)
1000 m	= 1 kilometre (km)

Short lengths can be measured in millimetres.

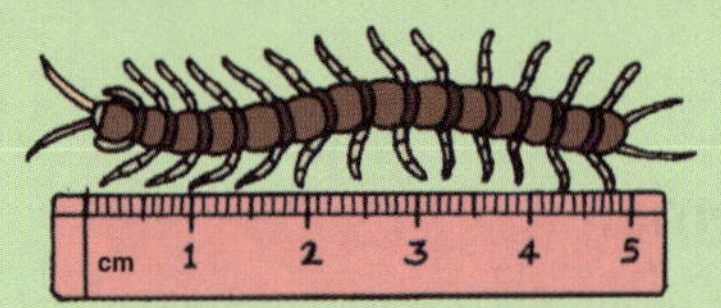

Long distances can be measured in kilometres.

PARIS 380km

I Write these equivalent lengths.

a $3\frac{1}{2}$ km = ☐ m

b 40 mm = ☐ cm

c 150 cm = ☐ m

d 8 cm = ☐ mm

e $\frac{1}{4}$ m = ☐ cm

f 6500 m = ☐ km

g 22 cm = ☐ mm

h 18 km = ☐ m

i $4\frac{3}{4}$ m = ☐ cm

j 65 mm = ☐ cm

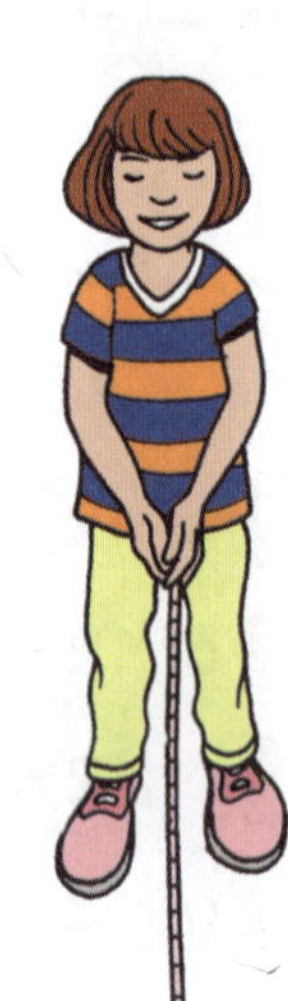

II Use a ruler to measure these lines in millimetres.

a ☐ mm

b ☐ mm

c ☐ mm

d ☐ mm

e ☐ mm

Multiplication and division

Multiplication and division are **linked**.

$6 \times 5 = 30$

If you know this, there are three other facts you also know.

$5 \times 6 = 30$
$30 \div 5 = 6$
$30 \div 6 = 5$

The three numbers 6, 5 and 30 are called a **trio**.

I Write four facts for each of these trios.

a 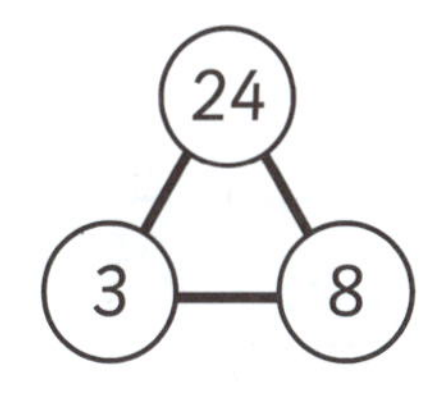

3	×	8	=	24
6	×	3	=	24
24	÷	8	=	3
24	÷	3	=	8

b 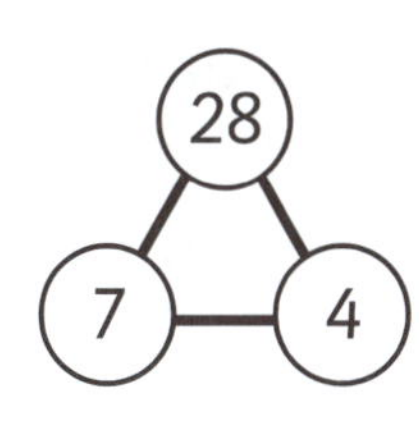

7	×	4	=	28
4	×	7	=	28
28	÷	4	=	7
28	÷	7	=	4

Write the missing numbers.

c [6] × 6 = 36

d 8 × [4] = 32

e [21] ÷ 3 = 7

f 45 ÷ [5] = 9

g 54 ÷ [6] = 9

h [24] ÷ 4 = 6

i 4 × [4] = 16

j [8] × 6 = 48

II If a number cannot be divided exactly, it leaves a remainder. Draw a line to join each division to its matching remainder.

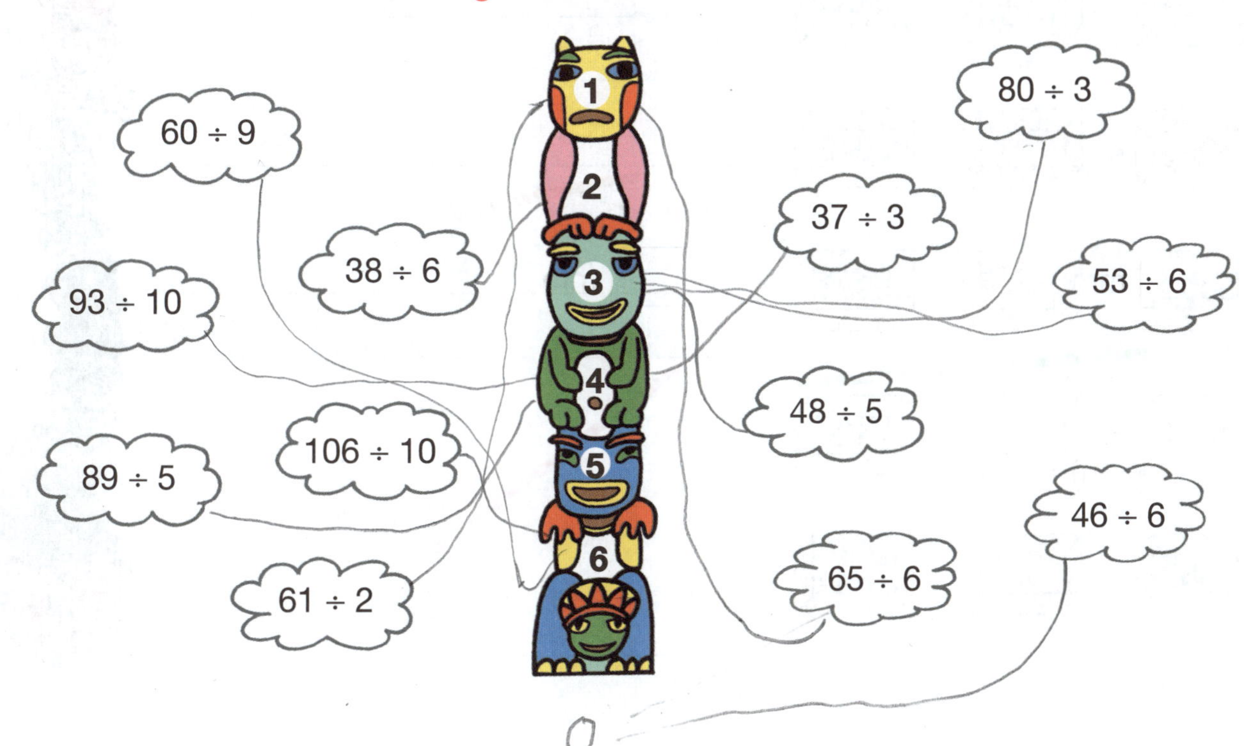

Comparing numbers

The symbols > and < are used to compare numbers.

means 'is less than'	means 'is greater than'
729 < 750	2500 > 2100
729 is less than 750	2500 is greater than 2100

I Write the signs > or < for each pair of numbers.

a	455	>	396	g	3750	>	3079
b	817	<	870	h	6002	<	6010
c	958	>	936	i	5299	<	5300
d	1904	<	2301	j	7451	>	7415
e	1850	>	1508	k	5306	<	5311
f	2001	>	1998	l	9038	>	9009

II Write the numbers that could go in each middle box.

a 4169 > 4166 > 4164   4169 4166 4164

b 3838 < 3839 < 3842   3838 3839 3842

c 9002 > ☐ > 8996   9002 ____ 8996

d  < 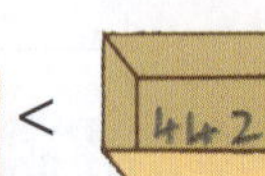 <    4421 4424 4426

e  >  >    7082 7081 7076

3-D shapes

A **polyhedron** is a 3-D shape with flat faces.

A cube is a polyhedron. It has:

- 8 corners (vertices)
- 12 edges
- 6 faces.

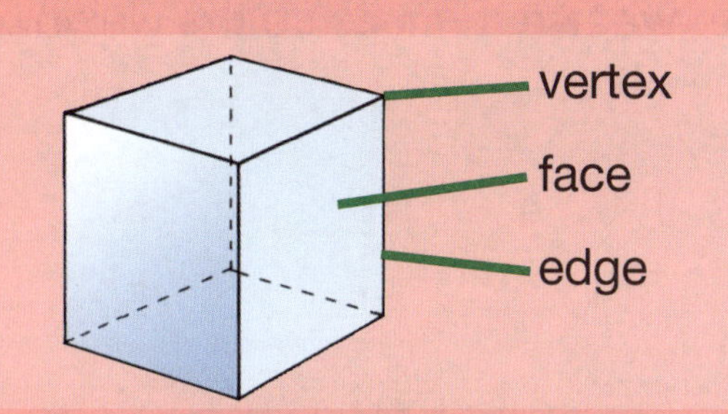

I Name each shape. Choose the correct word from the box.

cylinder
cone
cube
pyramid
sphere
cuboid

a 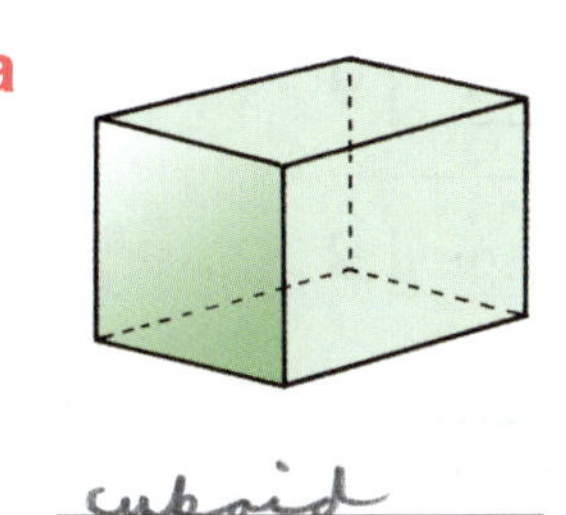 cuboid

c 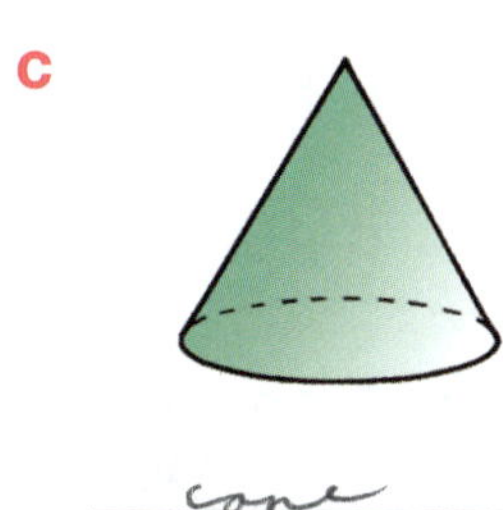 cone

e 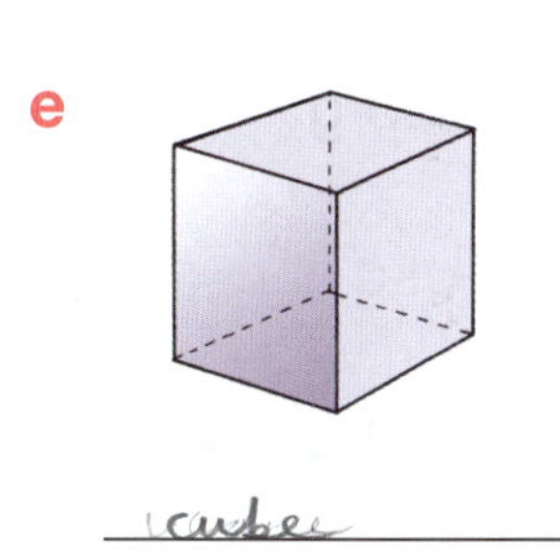 cube

b 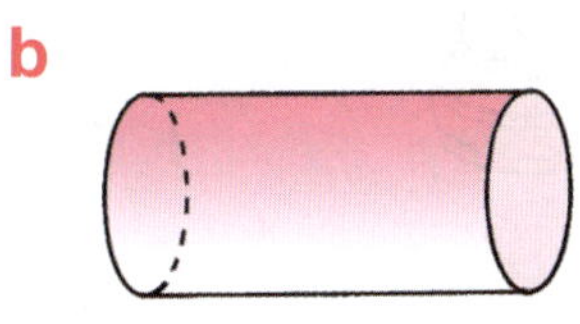 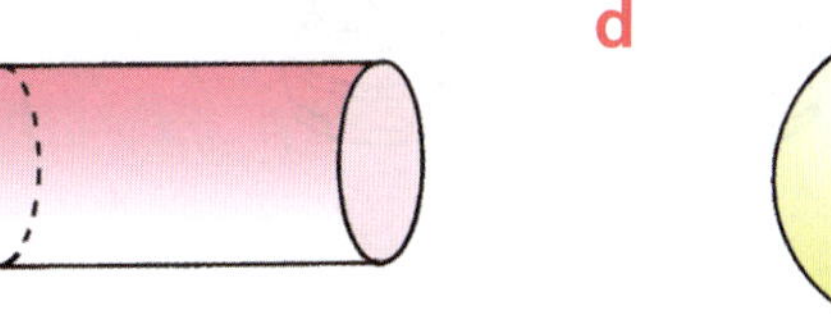 cylinder

d  sphere

f 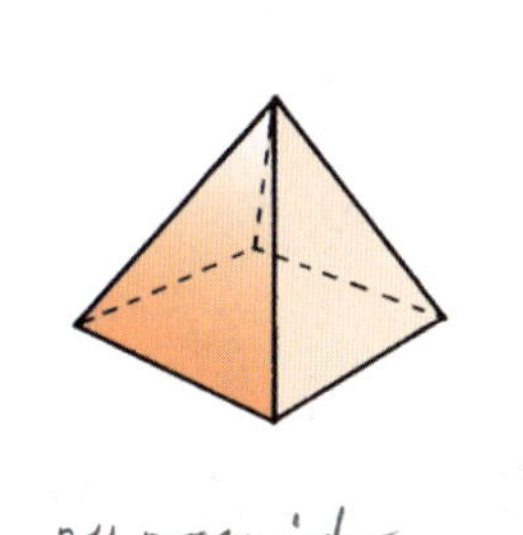 pyramid

II Write how many faces, edges and vertices each shape has.

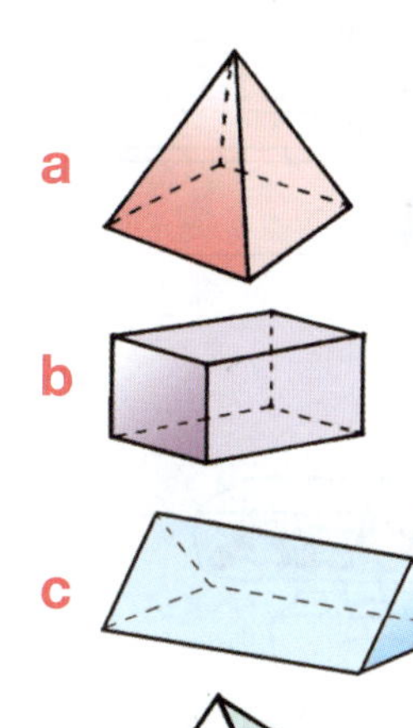

	faces	edges	vertices
a	5	8	5
b	6	12	8
c	5	9	6
d	4	5	4

Measuring mass

Kilograms (kg) and **grams** (g) are some of the units we use to measure the weight or mass of an object.

$1000\text{ g} = 1\text{ kg}$ $\quad$ $250\text{ g} = \frac{1}{4}\text{ kg}$

$500\text{ g} = \frac{1}{2}\text{ kg}$ $\quad$ $750\text{ g} = \frac{3}{4}\text{ kg}$

I Write these equivalent units.

a 2000 g = ☐ kg

b $1\frac{1}{2}$ kg = ☐ g

c 5500 g = ☐ kg

d 1250 g = ☐ kg

e 7 kg = ☐ g

f $3\frac{1}{4}$ kg = ☐ g

g 10 kg = ☐ g

h 6750 g = ☐ kg

i $2\frac{1}{2}$ kg = ☐ g

j $4\frac{3}{4}$ kg = ☐ g

k 9500 g = ☐ kg

l $1\frac{3}{4}$ kg = ☐ g

II Look at these scales. Write the mass shown in kilograms.

a

______

b

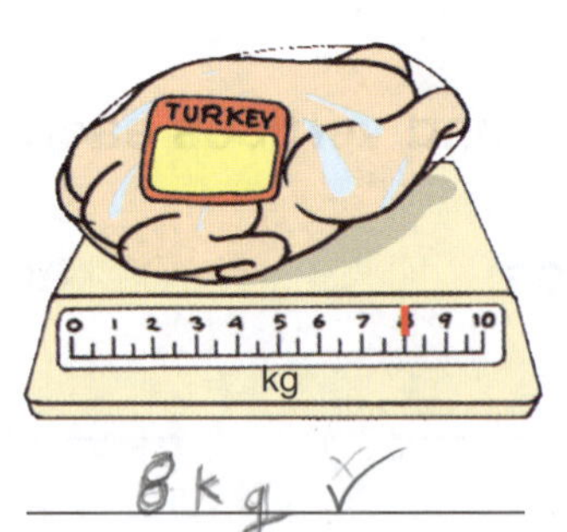

______

c

______

Write the mass shown in grams.

d

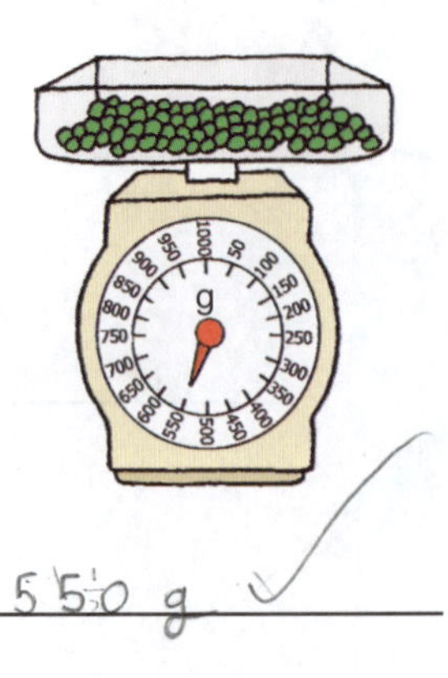

______

e

______

f

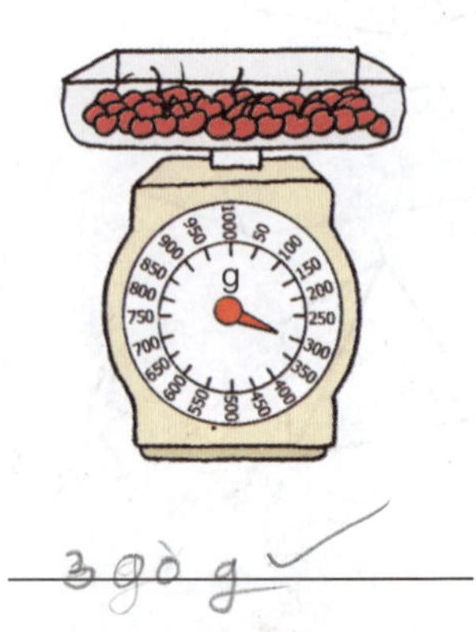

______

Addition

When you add numbers, decide whether to use a **mental method**, or whether you need to use the **written method**.

Mental method 53 + 48

Example

53 add 50 is 103
Take away 2 is 101

53 add 40 is 93
93 add 8 is 101

Written method 156 + 75

Example

$$\begin{array}{r} 156 \\ +\ 75 \\ \hline 231 \\ \hline {}_{1\ 1\ } \end{array}$$

Add the ones (6 + 5)

Then the tens (50 + 70 + 10)

Then the hundreds (100 + 100)

I Use your own methods to add these. Colour the star if you used a mental method.

a 51 + 43 = 94

b 38 + 63 = 101

c 29 + 35 = 64

d 86 + 62 = 158

e 91 + 74 = 165

f 57 + 69 = 126

g 37 + 94 = 131

h 75 + 66 = 141

i 88 + 83 = 171

j 124 + 132 = 256

k 146 + 105 = 251

l 135 + 166 = 301

II Answer these.

a $\begin{array}{r} 386 \\ +\ 58 \\ \hline 444 \end{array}$

b $\begin{array}{r} 274 \\ +\ 81 \\ \hline 355 \end{array}$

c $\begin{array}{r} 546 \\ +\ 74 \\ \hline 620 \end{array}$

d $\begin{array}{r} 914 \\ +\ 87 \\ \hline 1001 \end{array}$

e $\begin{array}{r} 676 \\ +\ 78 \\ \hline 754 \end{array}$

f $\begin{array}{r} 727 \\ +\ 83 \\ \hline 810 \end{array}$

Area

The area of a shape can be found by **counting squares on a grid.**

Count half squares for shapes with straight sides.

For irregular shapes, count the squares that are covered more than half.

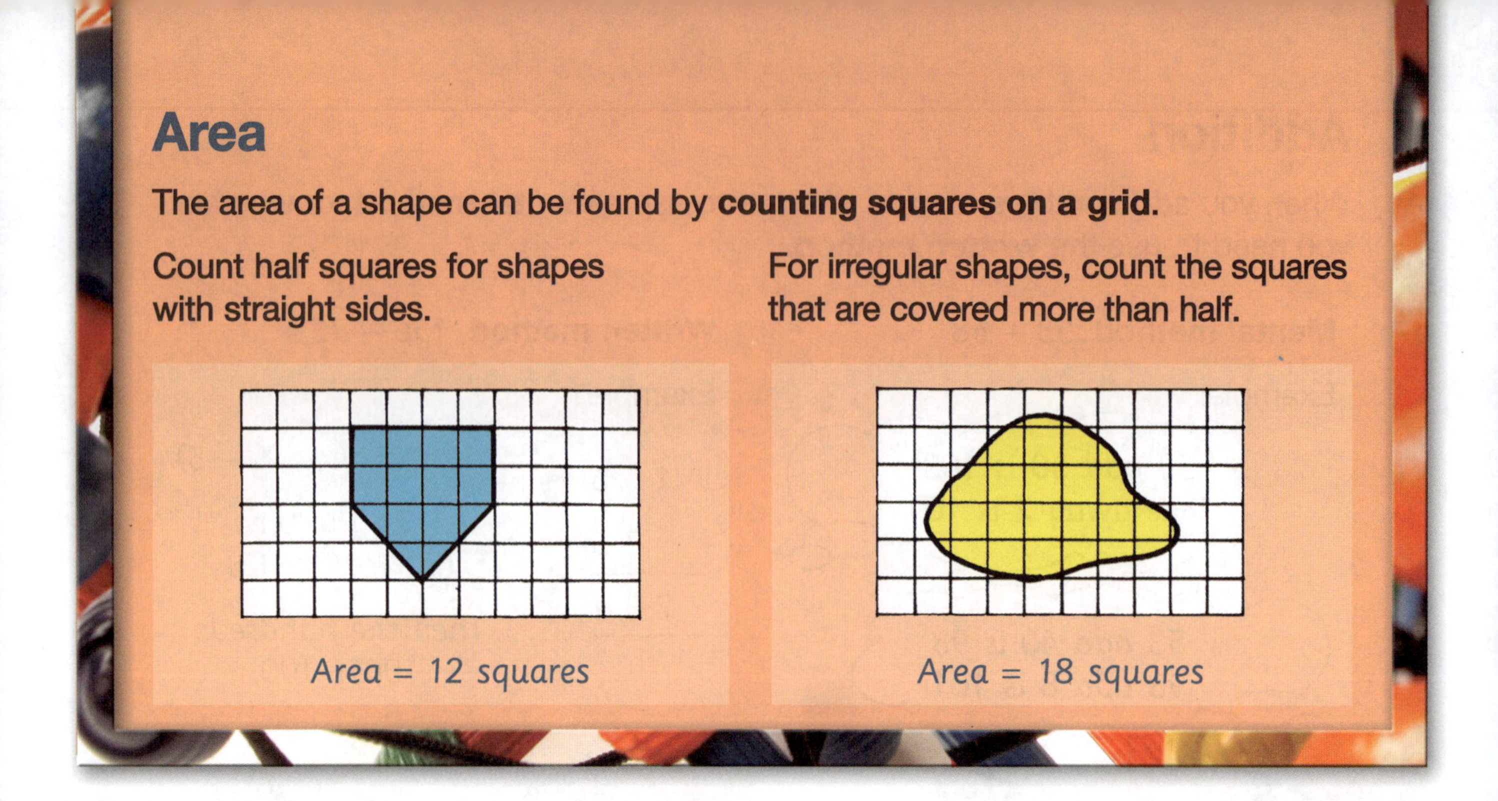

I **Work out the area of each shape.**

a Area = 10 squares

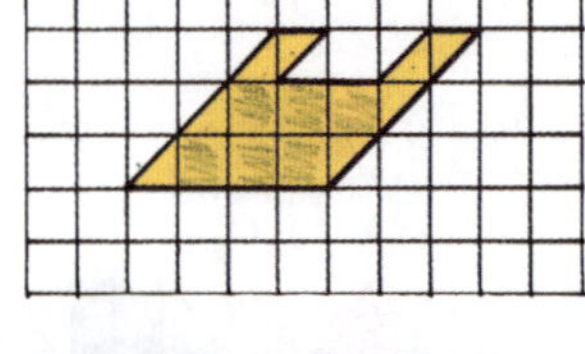

b Area = 21 squares

c Area = 16 squares

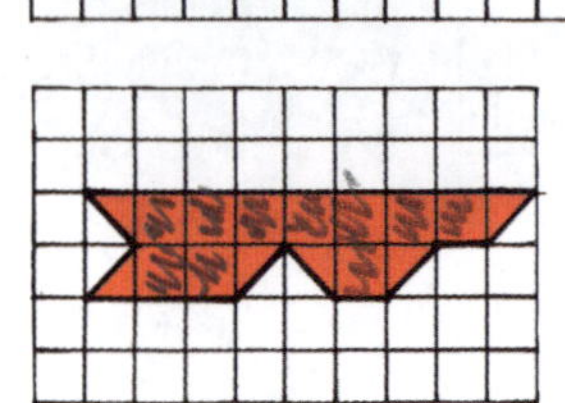

Work out the approximate area of these shapes.

d Area = 41 squares

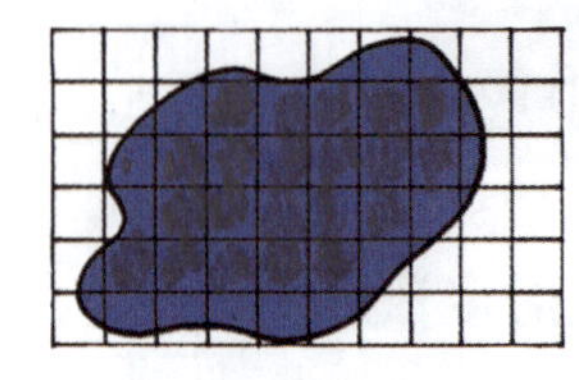

e Area = 23 squares

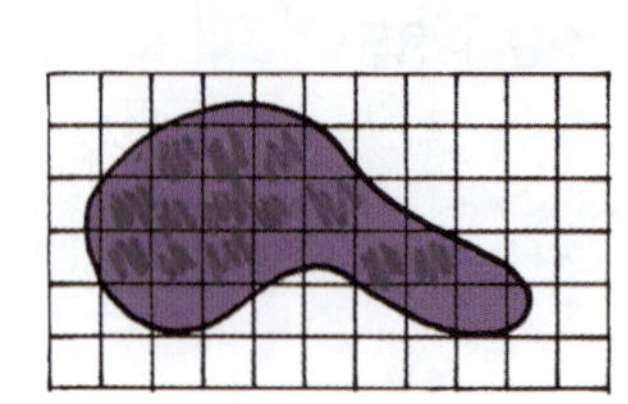

II **Draw three different shapes with an area of 8 squares.**

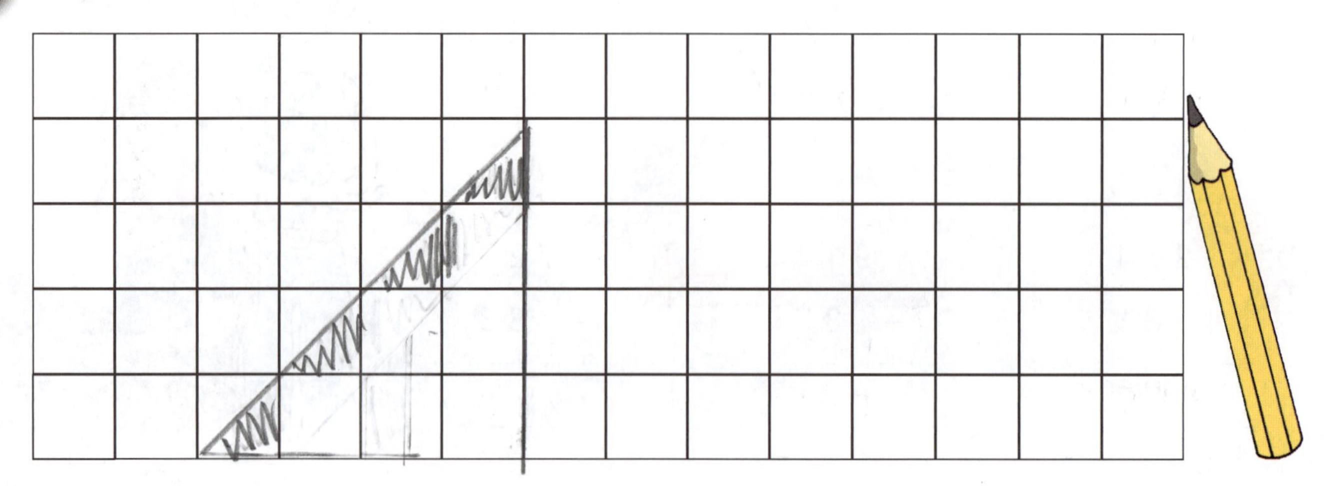

Measuring perimeter

The perimeter of a shape is the **distance all around the edge.**

The perimeter of this triangle is
3 cm + 4 cm + 5 cm = 12 cm

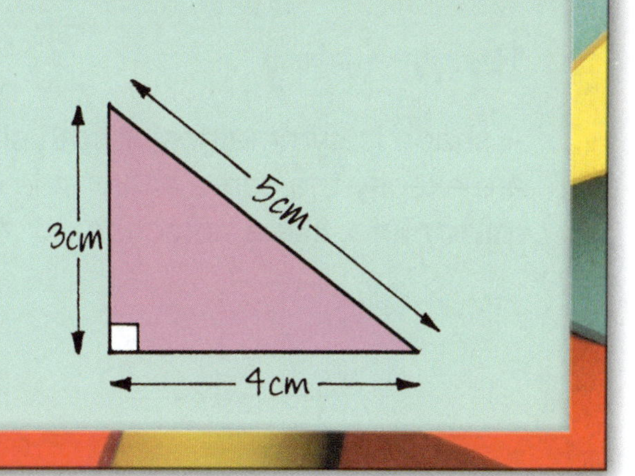

I Write the perimeter of each rectangle.

a Perimeter: 16 cm

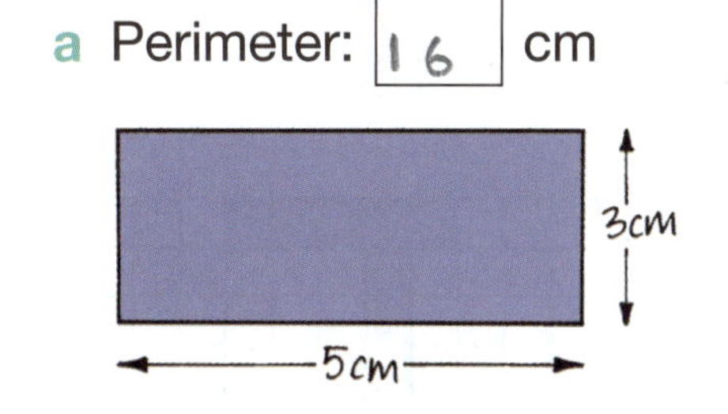

b Perimeter: 14 cm

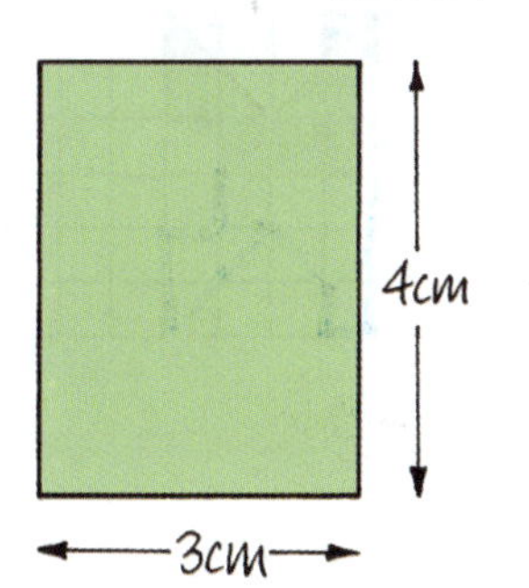

c Perimeter: 20 cm

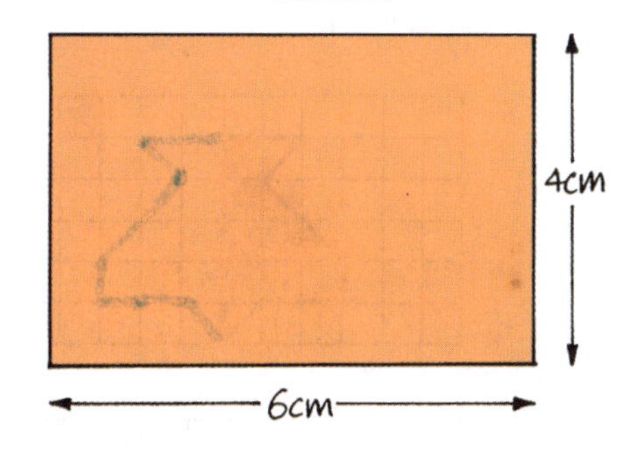

d Perimeter: 20 cm

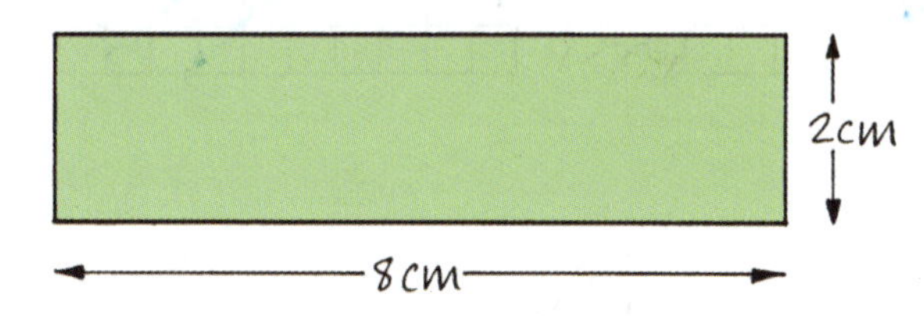

e Perimeter: 50 cm

5cm
2cm

II Use a ruler to measure the perimeter of each shape.

a

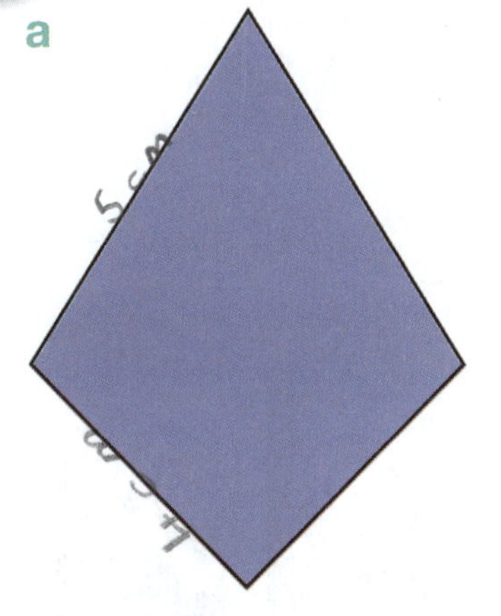

Perimeter = 18 cm

b

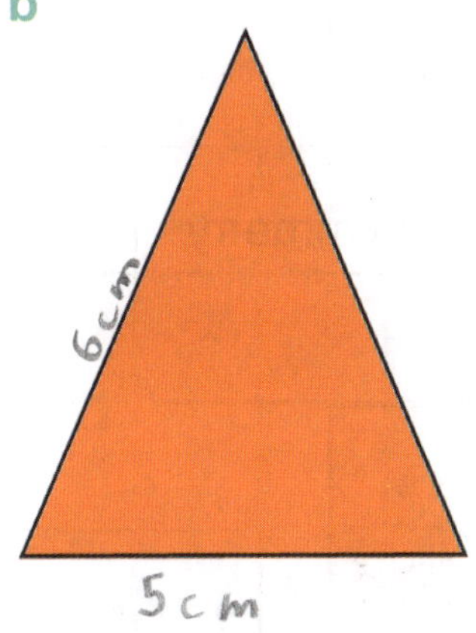

Perimeter = 22 cm

c

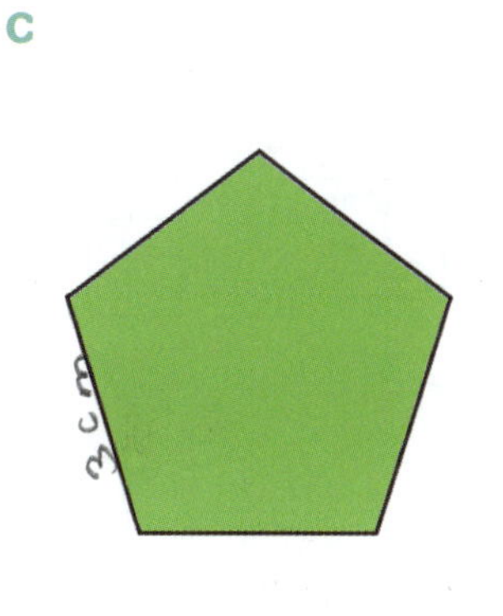

Perimeter = 16 cm

d

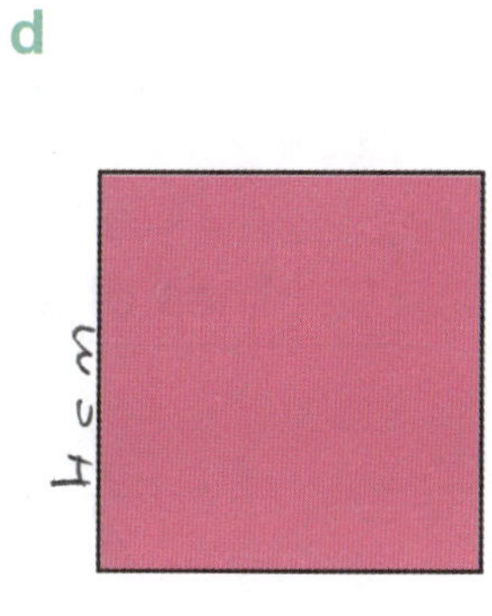

Perimeter = 16 cm

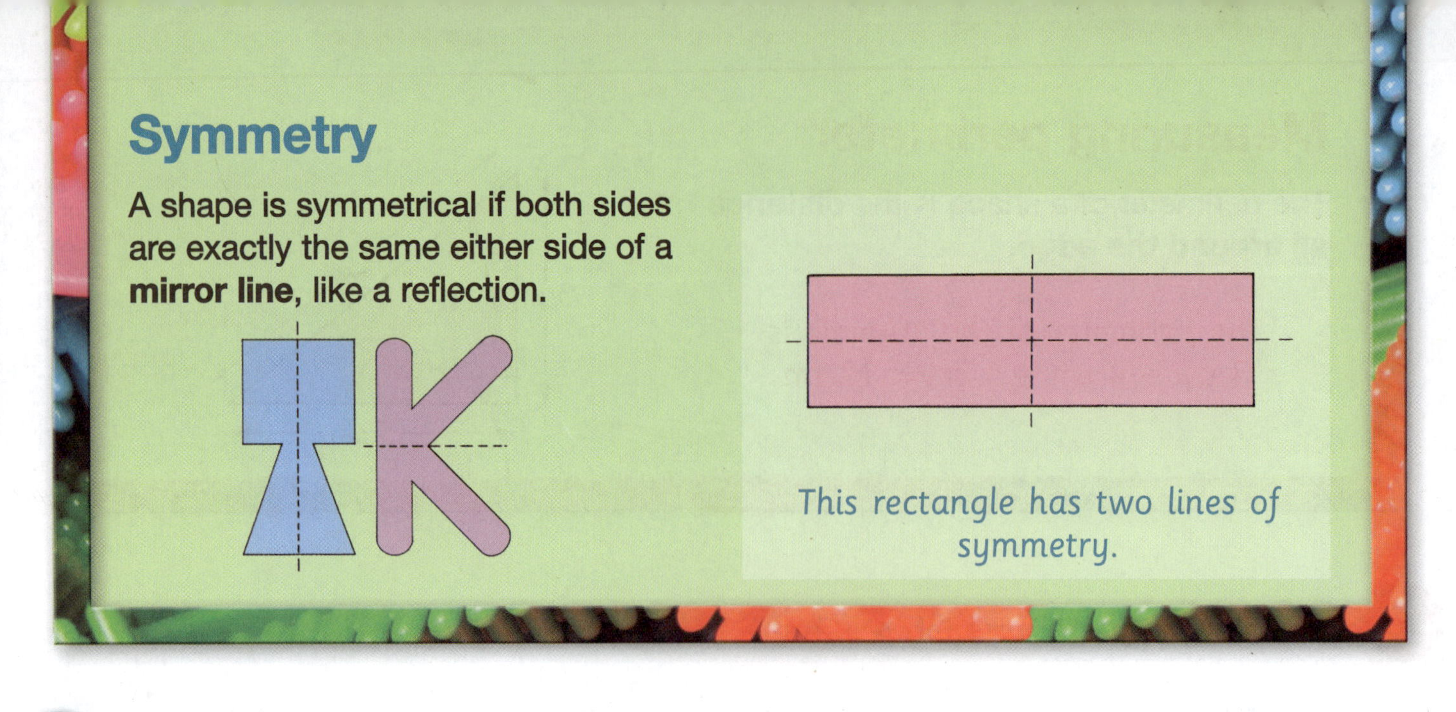

Symmetry

A shape is symmetrical if both sides are exactly the same either side of a **mirror line**, like a reflection.

This rectangle has two lines of symmetry.

I Draw the reflection of each of these.

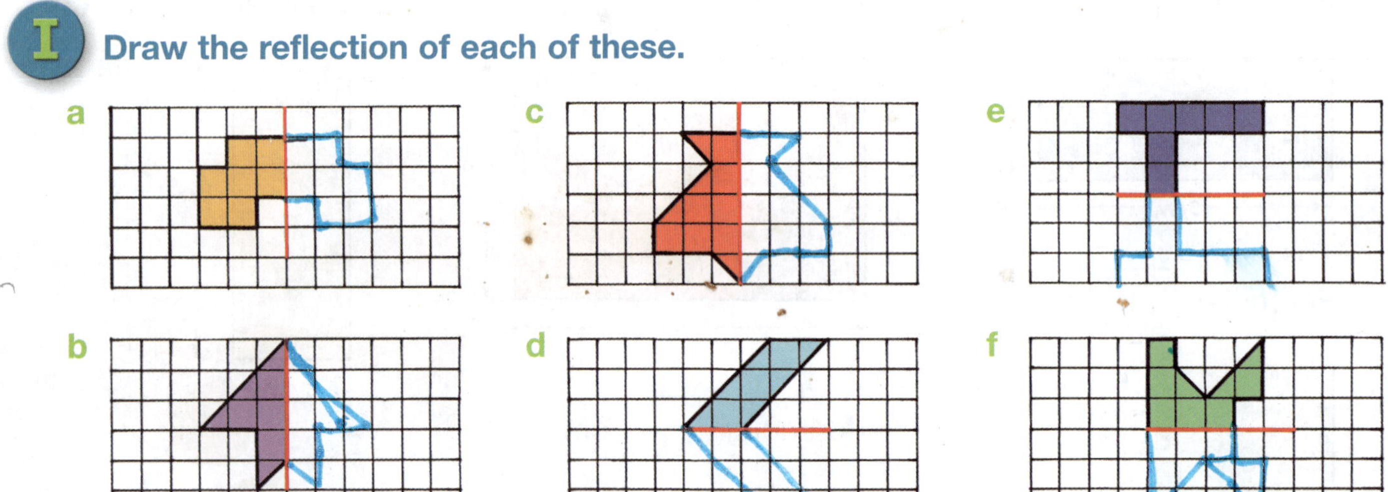

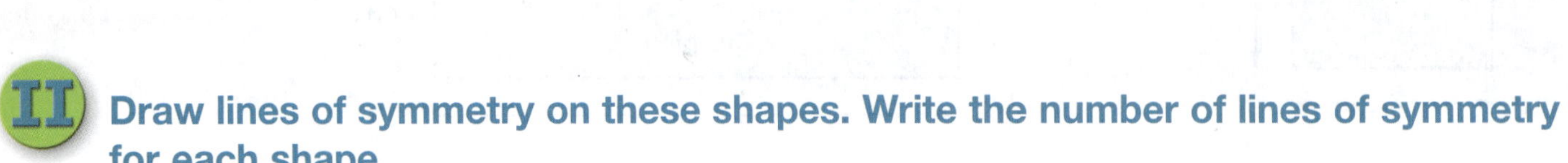

II Draw lines of symmetry on these shapes. Write the number of lines of symmetry for each shape.

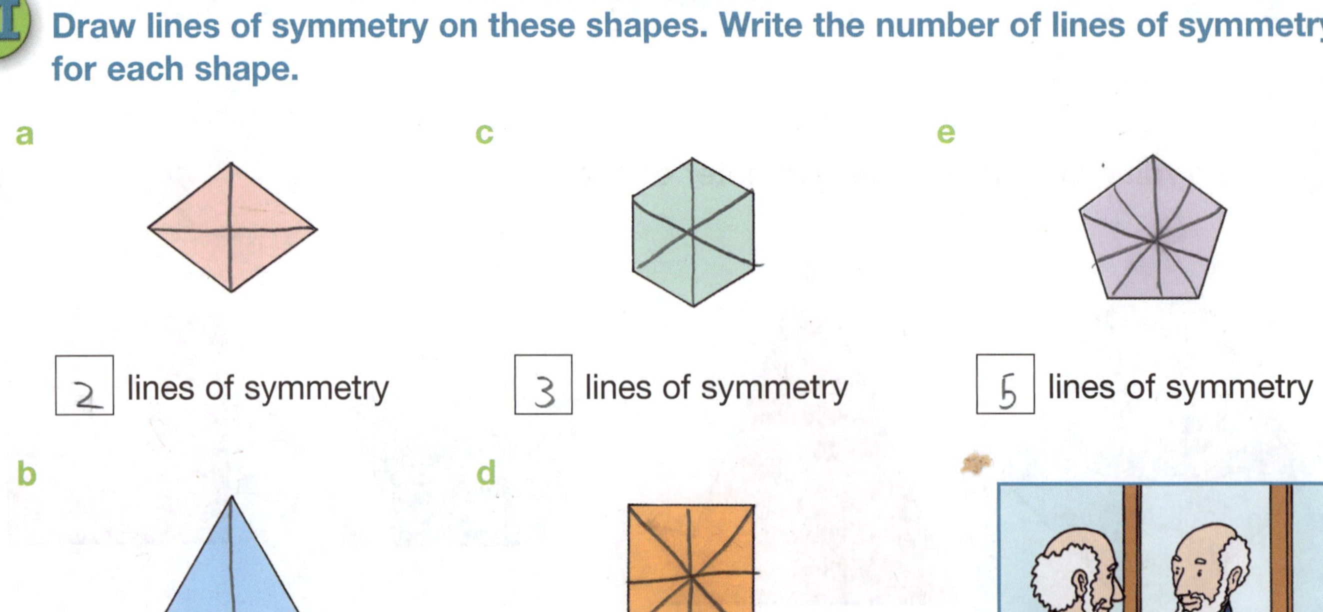

Measuring capacity

Litres (l) and **millilitres** (ml) are some of the units we use to measure the capacity of liquids in containers.

1000 ml = 1 l	750 ml = $\frac{3}{4}$ l
250 ml = $\frac{1}{4}$ l	500 ml = $\frac{1}{2}$ l

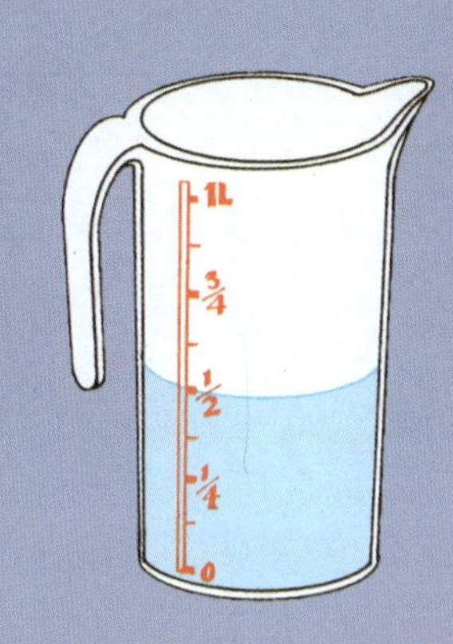

I Write the equivalent units.

a 3000 ml = 3 l

b $1\frac{1}{2}$ l = 1500 ml

c 6 l = 6000 ml

d 2250 ml = 225 l

e 1750 ml = 175 l

f 10 l = 10000 ml

g $3\frac{1}{2}$ l = 3500 ml

h 2000 ml = 2 l

i 4500 ml = 45 l

j $8\frac{3}{4}$ l = 8750 ml

k 5750 ml = 575 l

l $6\frac{3}{4}$ l = 6750 ml

II Write the capacity each jug shows in millilitres.

a

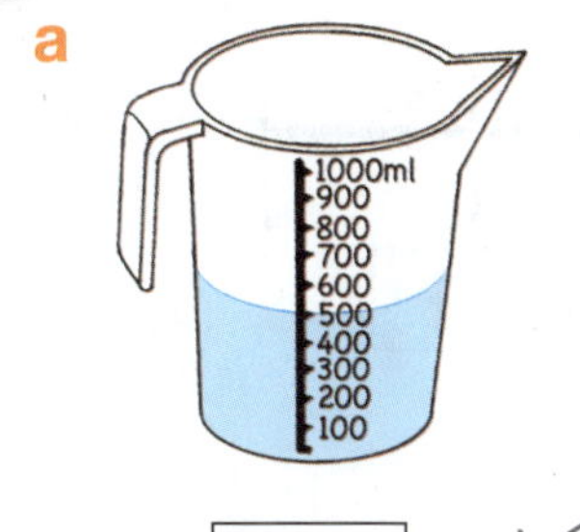

500 ml

b

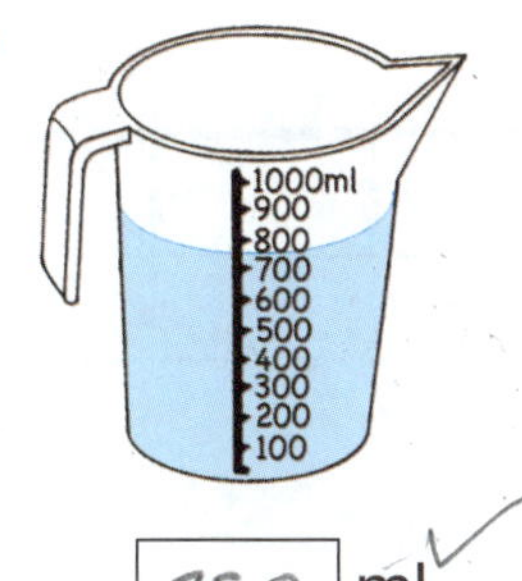

750 ml

c

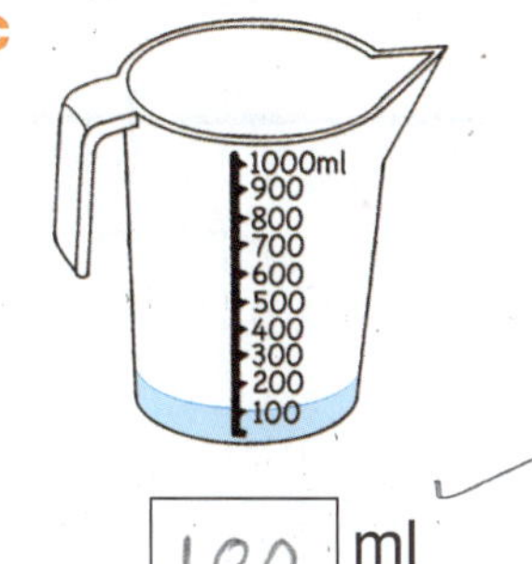

100 ml

d

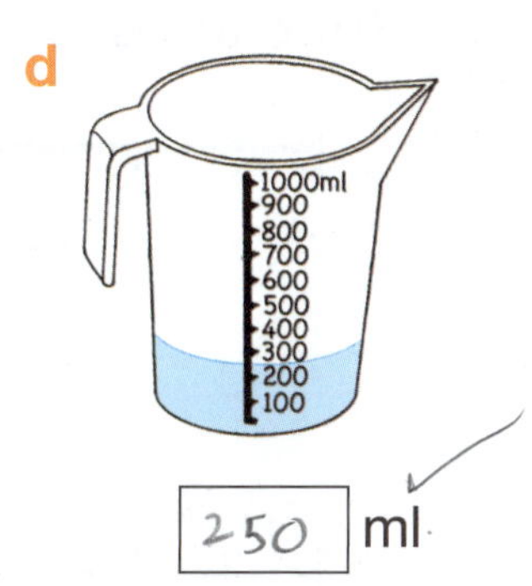

250 ml

e

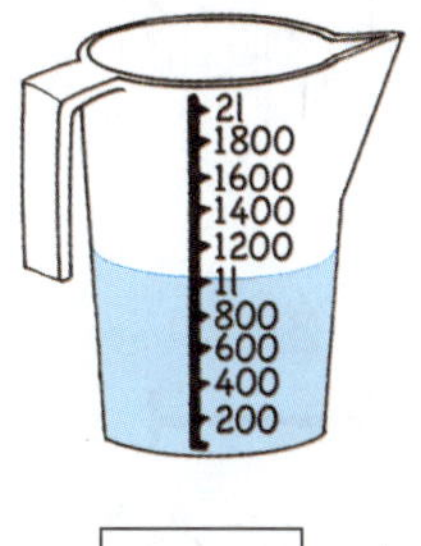

1000 ml

f

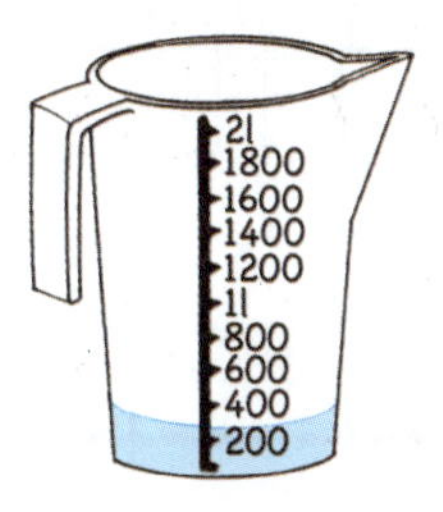

250 ml

g

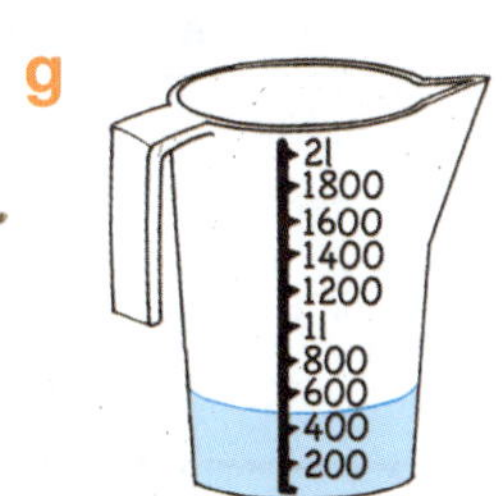

500 ml

h

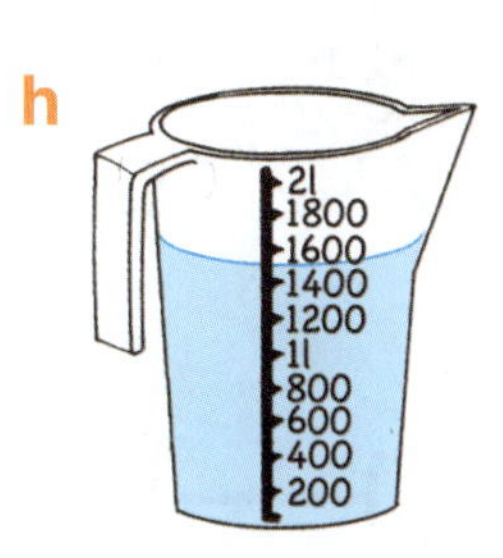

1500 ml

Decimals

A **decimal point** is used to separate whole numbers from fractions.

$0.1 = \frac{1}{10}$

$0.5 = \frac{1}{2}$

$1.2 = 1\frac{2}{10}$

$3.9 = 3\frac{9}{10}$

$45.6 = 40 + 5 + \frac{6}{10}$

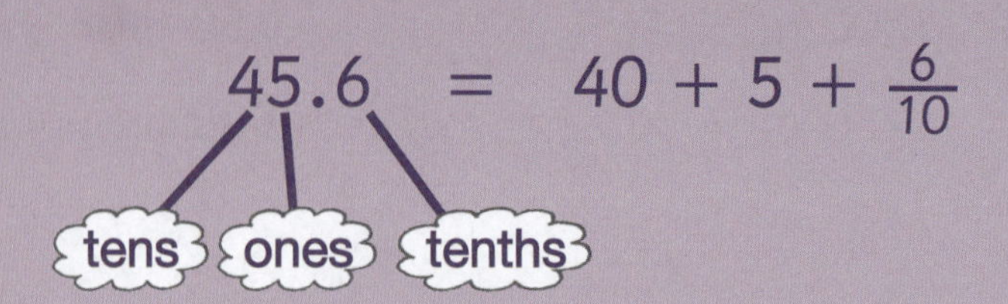

I Change these fractions to decimals.

a $\frac{3}{10}$ = 0.3

b $\frac{1}{2}$ = 0.5

c $\frac{2}{10}$ = 0.2

d $\frac{7}{10}$ = 0.7

e $1\frac{7}{10}$ = 1.7

f $3\frac{3}{10}$ = 3.3

g $4\frac{1}{2}$ = 4.5

h $7\frac{9}{10}$ = 7.9

i $42\frac{1}{2}$ = 42.5

j $19\frac{4}{10}$ = 19.4

k $68\frac{8}{10}$ = 68.8

l $59\frac{6}{10}$ = 59.6

II Write the decimals on these number lines.

a

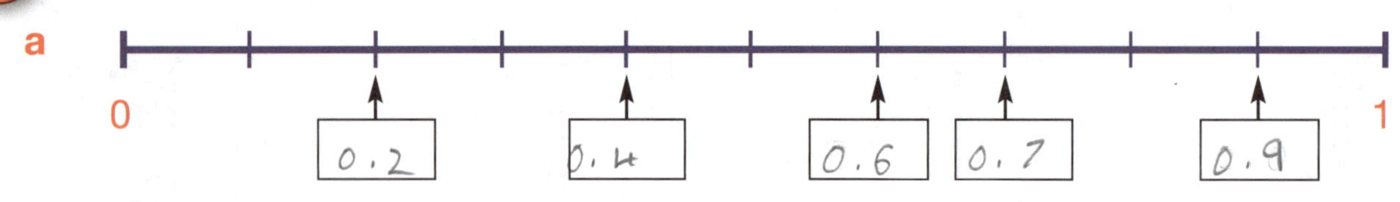

b

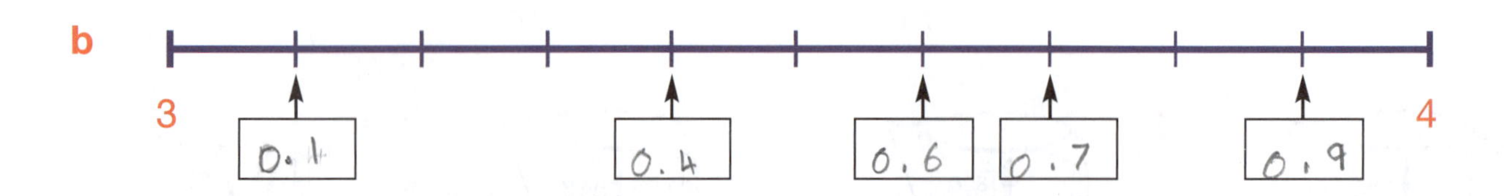

c

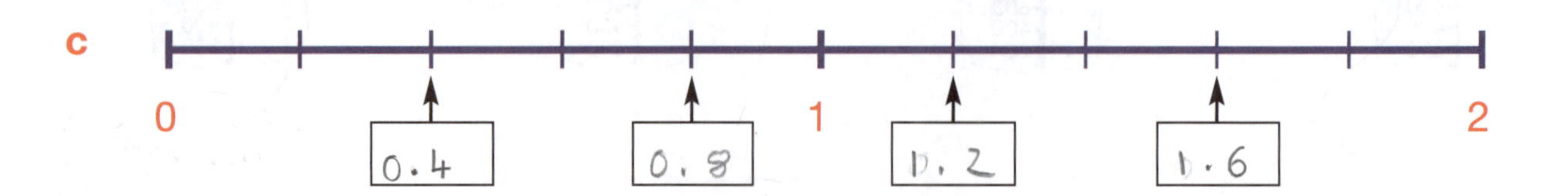

Reading pictograms

Pictograms are pictures or symbols to **show amounts.**

Check the number that each single picture represents.

This pictogram shows the number of skipping rope jumps in one minute by a group of children.

Jo	
Sam	
Alex	
Kim	
Ashley	

= 5 jumps   = 1, 2, 3 or 4 jumps

I Look at the skipping pictogram and answer these questions.

a How many skips did Sam jump? ____________

b Who jumped the most in one minute? ____________

c Who jumped a total of 37 skips? ____________

d How many more skips did Ashley jump than Sam? ____________

e Who jumped 12 fewer skips than Jo? ____________

f If Jo jumped 32 skips, how many more is this than Sam? ____________

g Write the number of skips for each child.

Jo: ☐ Sam: ☐ Alex: ☐ Kim: ☐ Ashley: ☐

II Carry out a 'skipping' survey. Ask family or friends to skip for a certain time and record the results as a pictogram. Decide on a symbol and a number this would represent.

name	number of skips

☐ = ☐ skips

Subtraction

When you subtract numbers, decide whether to use a **mental method**, or whether you need to use the **written method**.

Mental method 92 – 57 = 35

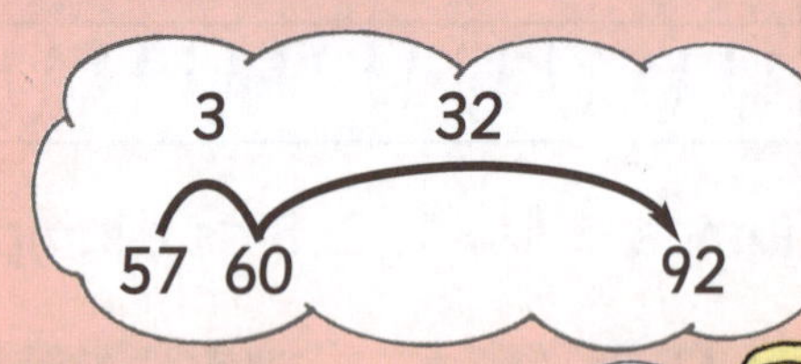

57 on to 60 is 3

60 on to 92 is 32

32 add 3 is 35

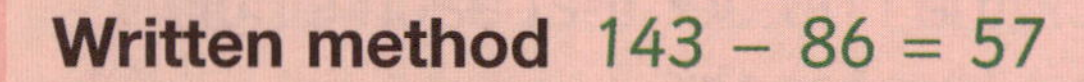

Written method 143 – 86 = 57

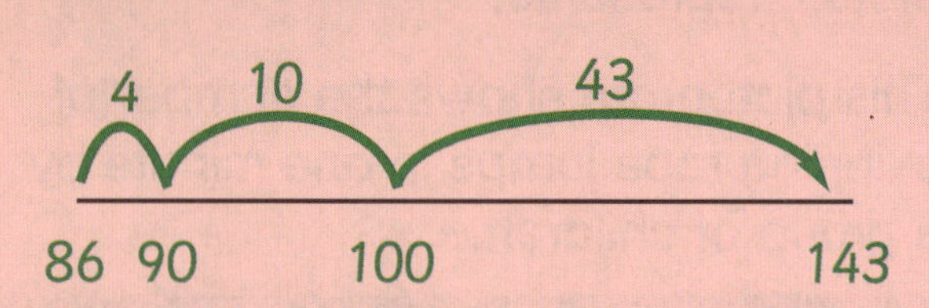

Count on from 86 in steps
4 + 10 + 43 = 57

These both use a number line to work out the answers.

I Use the number line method for these.

a 74 – 38 = [36]

38 ——— 74

b 81 – 46 = [43]

46 ——— 81

c 93 – 57 = [42]

57 ——— 93

d 125 – 87 = [38]

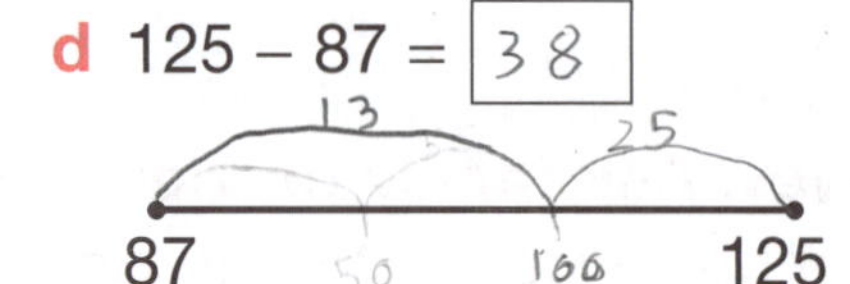

e 152 – 76 = [76]

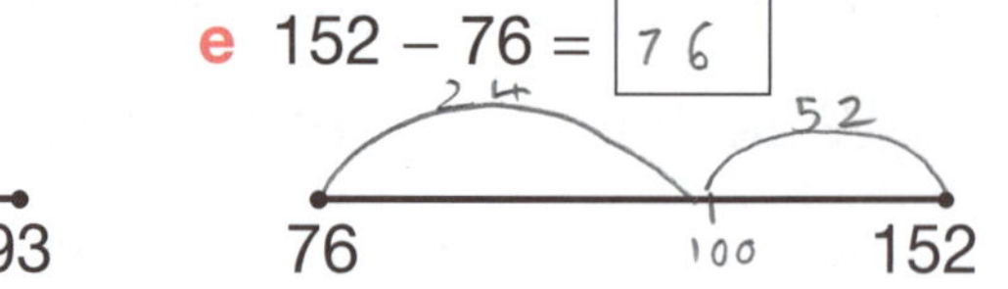

f 164 – 95 = [69]

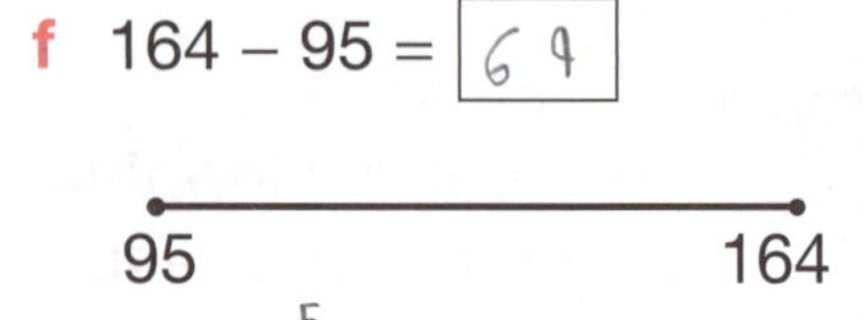

II Choose a method to work out the differences between these pairs of weights.

a

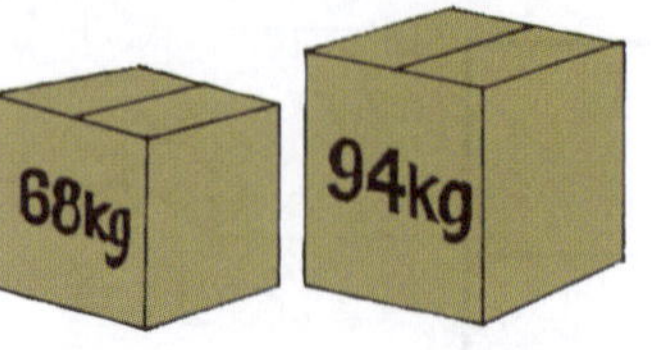
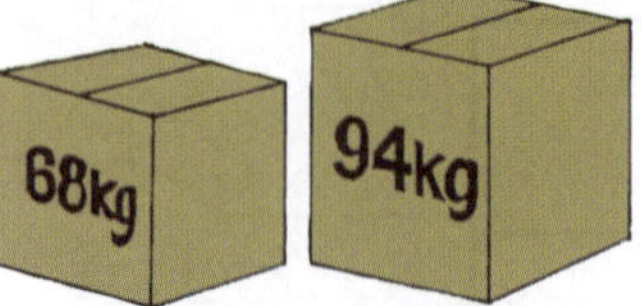

Difference: [22] kg

b

Difference: [48] kg

c

Difference: [58] kg

d

Difference: [46] kg

e

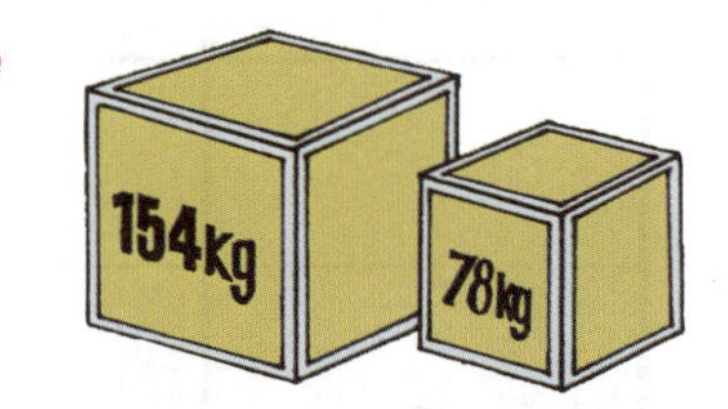

Difference: [76] kg

f

Difference: [77] kg

Equivalent fractions

Fractions that have the **same value** are called equivalent fractions.

$\frac{5}{10}$ is the same as $\frac{1}{2}$

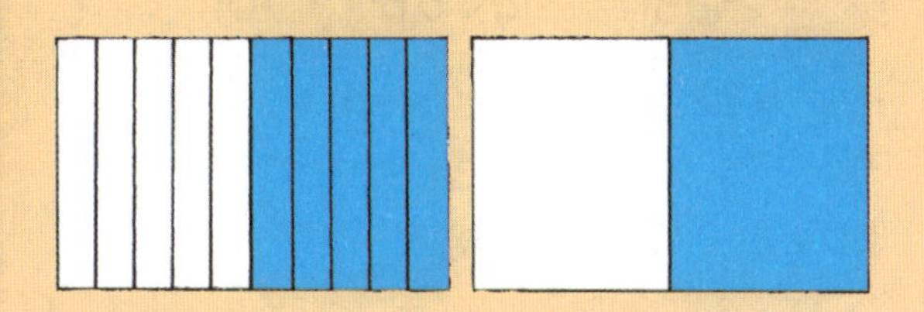

$\frac{1}{3}$ is the same as $\frac{2}{6}$

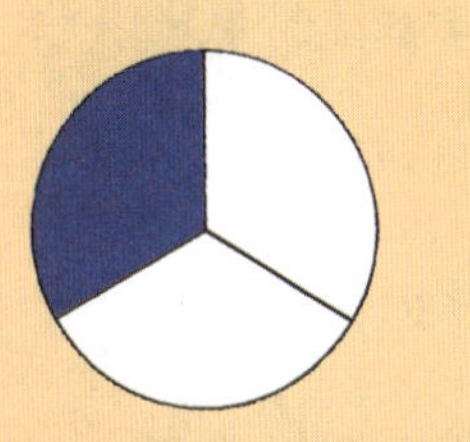

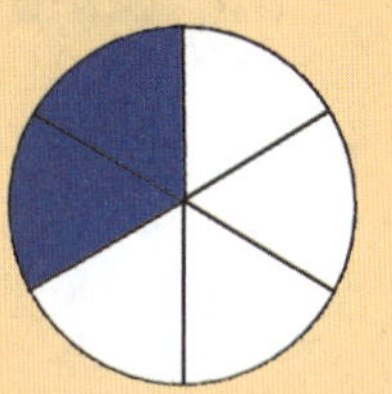

I Complete the equivalent fractions.

a 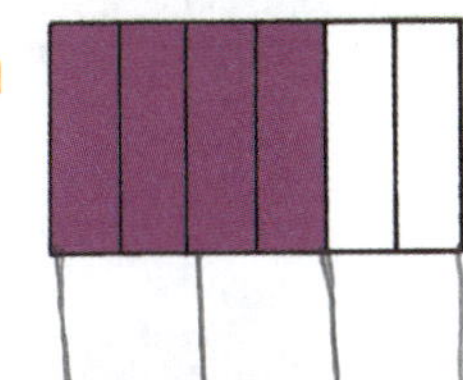

$\frac{4}{6} = \frac{2}{3}$

b 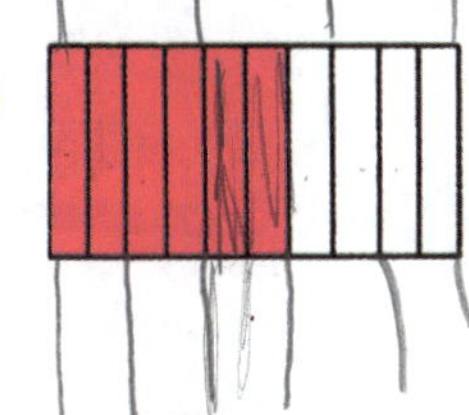

$\frac{6}{10} = \frac{3}{5}$

c 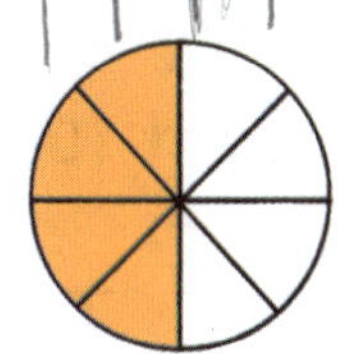

$\frac{4}{8} = \frac{1}{2}$

d 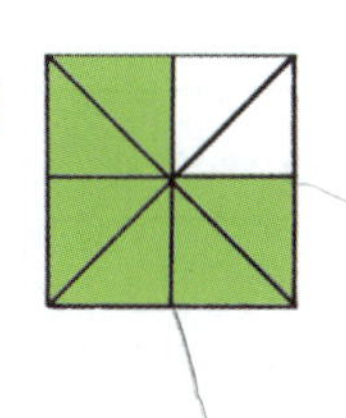

$\frac{6}{8} = \frac{3}{4}$

e $\frac{1}{3} = \frac{2}{8}$

f $\frac{4}{12} \; \frac{2}{6} = \frac{1}{3}$

g $\frac{5}{10} = \frac{1}{2}$

h $\frac{2}{5} = \frac{4}{10}$

i $\frac{6}{12} = \frac{1}{2}$

j $\frac{3}{4} = \frac{9}{12}$

II Cross out the fraction that is not equivalent to the others in each set.

a $\frac{1}{2}$ →

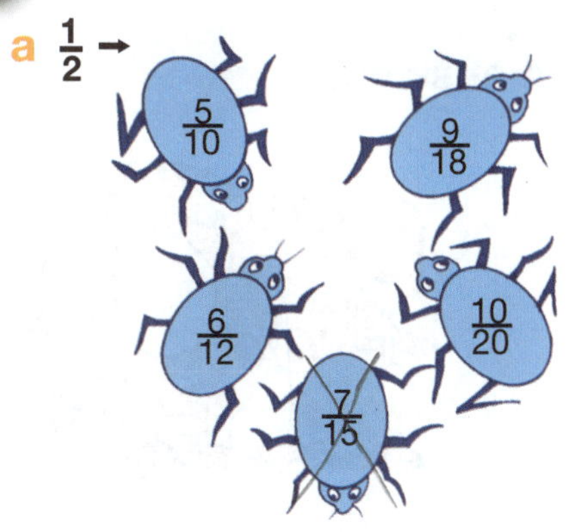

b $\frac{1}{4}$ →

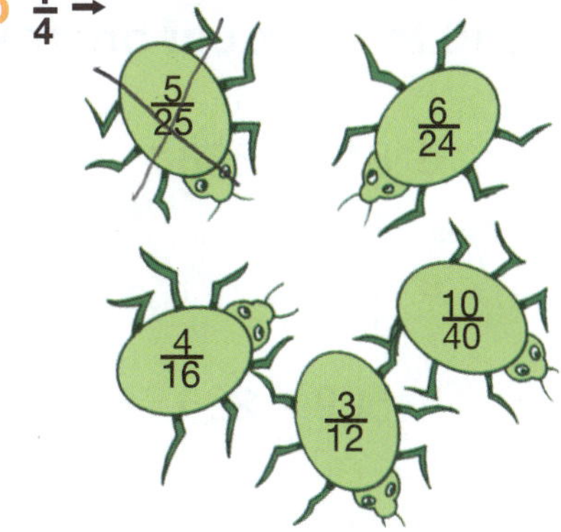

c $\frac{1}{3}$ →

3/9
6/18
4/12
10/30
8/27

Rounding numbers

Rounding to the nearest 10

38 rounds **up** to 40

214 rounds **down** to 210

Look at the **ones** digit.

- If it is 5 or more, round up to the next tens.
- If it is less than 5, the tens digit stays the same.

Rounding to the nearest 100

653 rounds **up** to 700

439 rounds **down** to 400

Look at the **tens** digit.

- If it is 5 or more, round up to the next hundreds.
- If it is less than 5, the hundreds digit stays the same.

This chart shows a list of some of the highest waterfalls in the world. Round each height to the nearest 10 m and 100 m.

Waterfall	Country	Total drop (m)	Rounded to the: Nearest 10 m	Nearest 100 m
Angel	Venezuela	979	980	1000
Tugela	South Africa	947	950	900
Mongefossen	Norway	774	770	700
Yosemite	USA	739	740	700
Tyssestrengane	Norway	646	650	700
Sutherland	New Zealand	581	580	600
Kjellfossen	Norway	561	560	600

Round these to the nearest 10 to work out approximate answers.

a 73 + 89 → 160 (70 90)

b 346 − 152 → 500 (300 200)

c 99 × 6 → 111 (100 10)

d 814 + 338 → 1110 (800 300)

e 17 × 9 → 30 (20 10)

f 509 − 296 → 800 (500 300)

Multiples

Multiples are like the numbers in the **times tables**.

Multiples of 2 are 2, 4, 6, 8, 10, 12 and so on.

Multiples of 5 are 5, 10, 15, 20, 25 and so on.

Multiples of a number do not come to an end at ×10, they go on and on.
For example 52, 98, 114, 230 are all multiples of 2.

I **Write these numbers in the correct boxes. Some of them will belong in more than one box.**

48 56 100 39 86 52 82 42 63 85 70 115 60 65

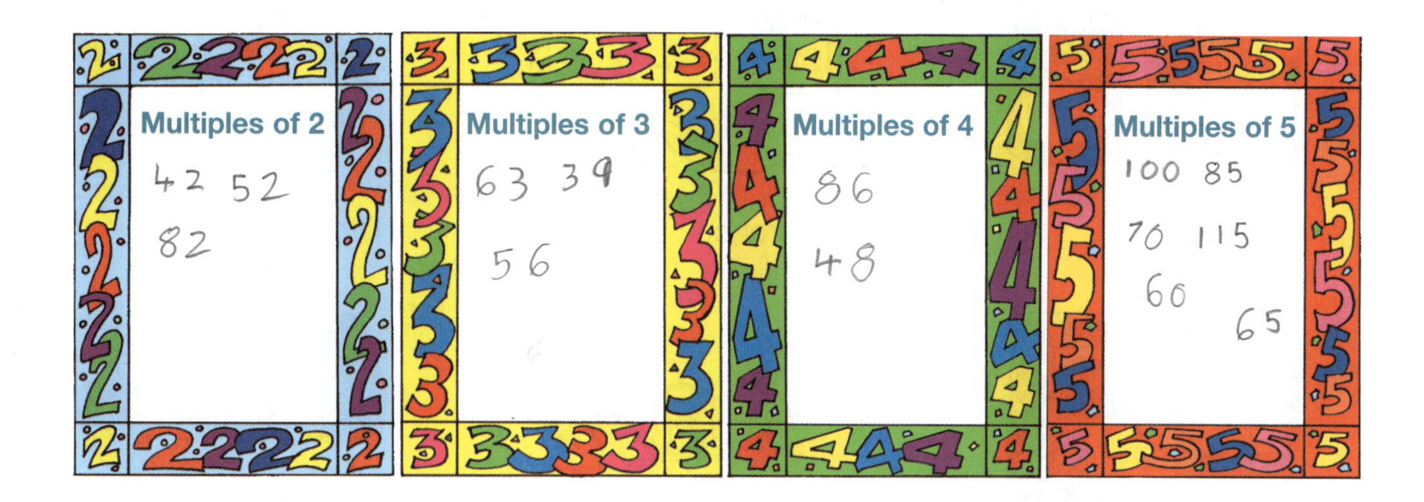

II **Colour all the multiples of 3 red.**
Colour all the multiples of 5 blue.
Look at the patterns on the grid.

1	2	3	4	5	6	7	8	9	10
11	12	13	14	15	16	17	18	19	20
21	22	23	24	25	26	27	28	29	30
31	32	33	34	35	36	37	38	39	40
41	42	43	44	45	46	47	48	49	50
51	52	53	54	55	56	57	58	59	60
61	62	63	64	65	66	67	68	69	70
71	72	73	74	75	76	77	78	79	80
81	82	83	84	85	86	87	88	89	90
91	92	93	94	95	96	97	98	99	100

Money problems

If you need to find the **difference** between two amounts, count on from the lower amount.

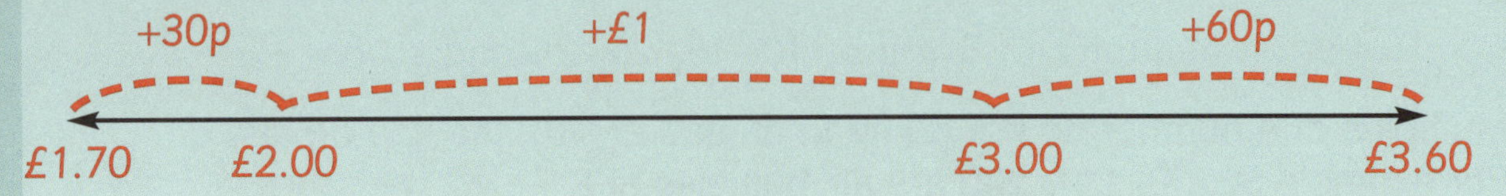

The difference between £1.70 and £3.60 is £1.90 (30p + £1 + 60p)

You can work out an amount of change in this way as well.

I Work out these differences.

a

Difference [ ]

c

Difference [ ]

e

Difference [ ]

b

Difference [ ]

d

Difference [ ]

f

Difference [ ]

II Draw a line to join these price labels to the correct change from £10.

a

b £8.99

c £7.89

d £3.69

e

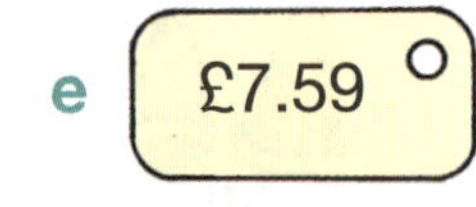

f

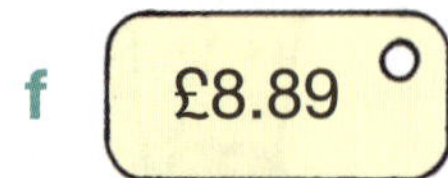

g

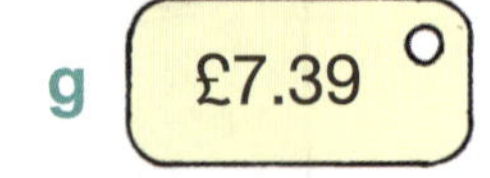

h £6.59

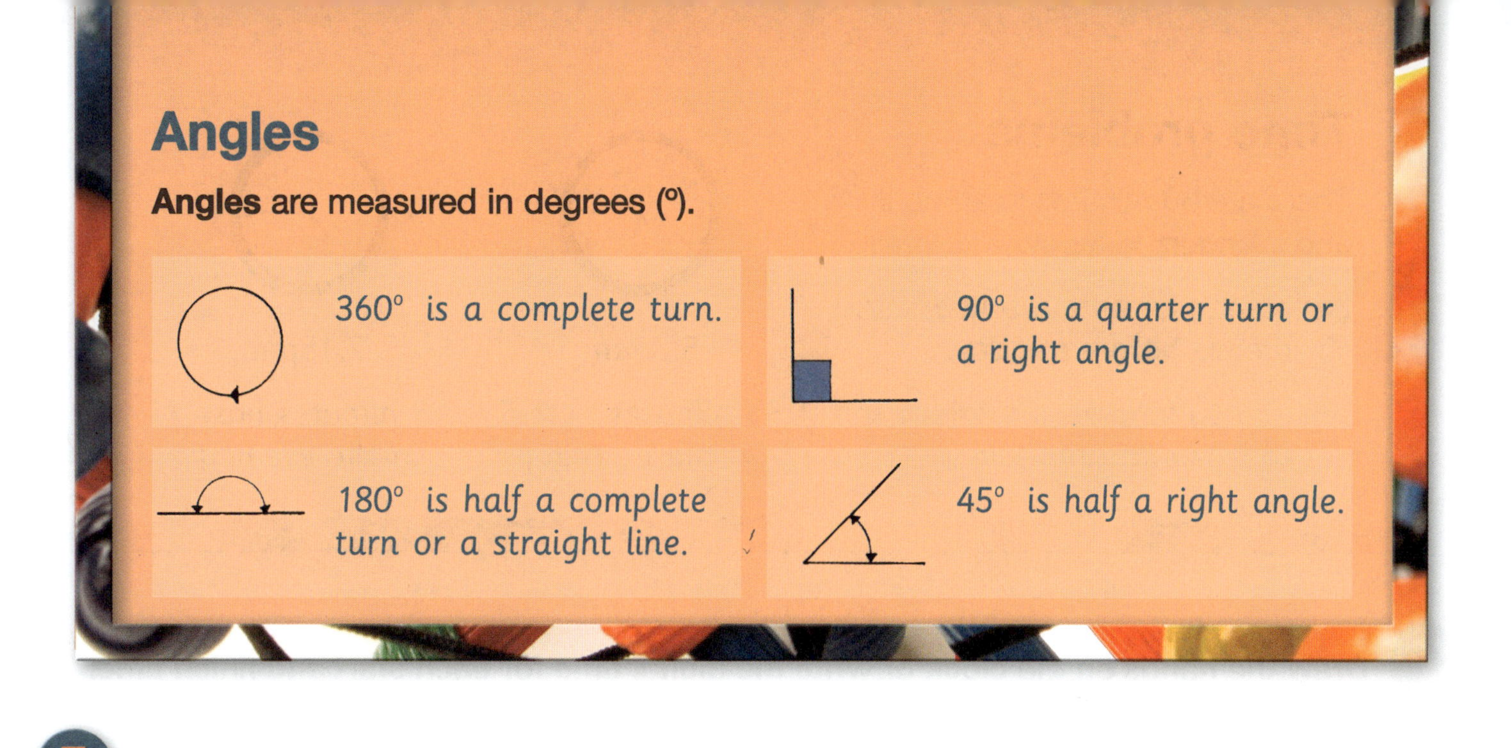

Angles

Angles are measured in degrees (°).

360° is a complete turn.

90° is a quarter turn or a right angle.

180° is half a complete turn or a straight line.

45° is half a right angle.

I **Tick the largest angle in each set. Is it greater or smaller than 90°? Circle the correct answer.**

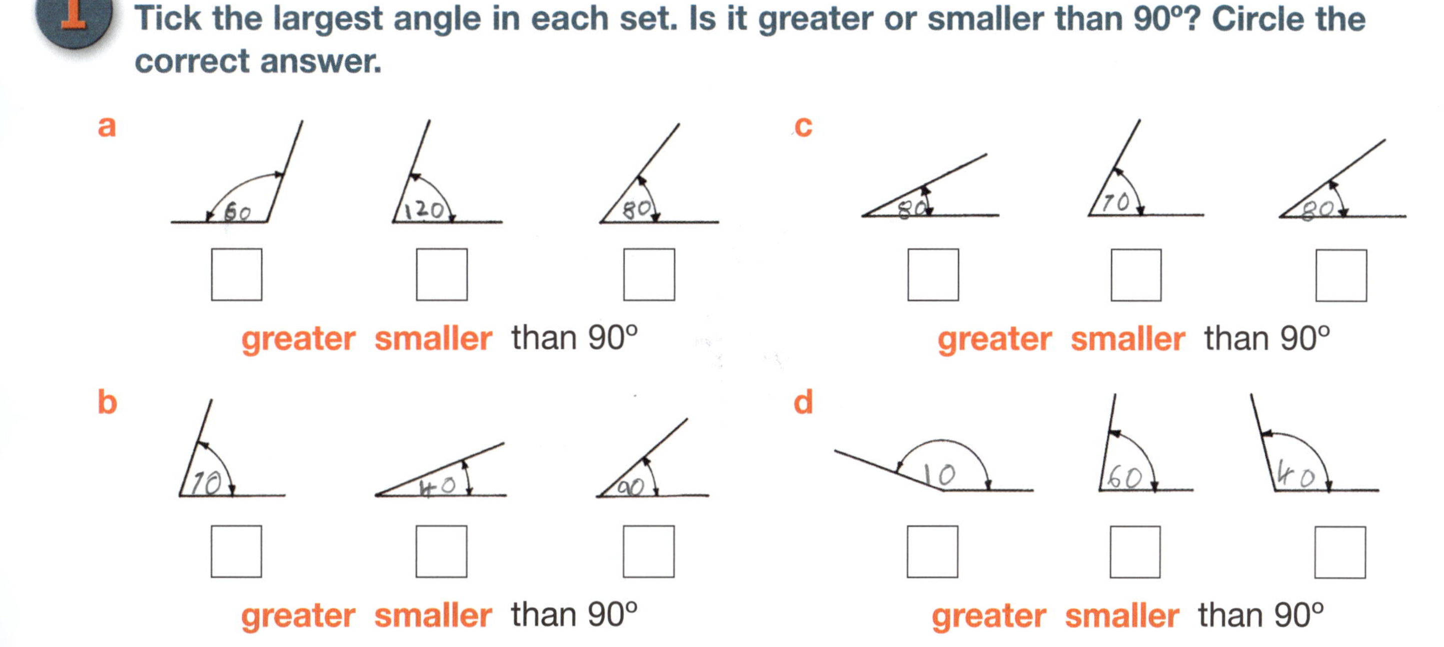

a greater smaller than 90°

b greater smaller than 90°

c greater smaller than 90°

d greater smaller than 90°

II **These are the eight compass directions. Write the direction you will face after turning.**

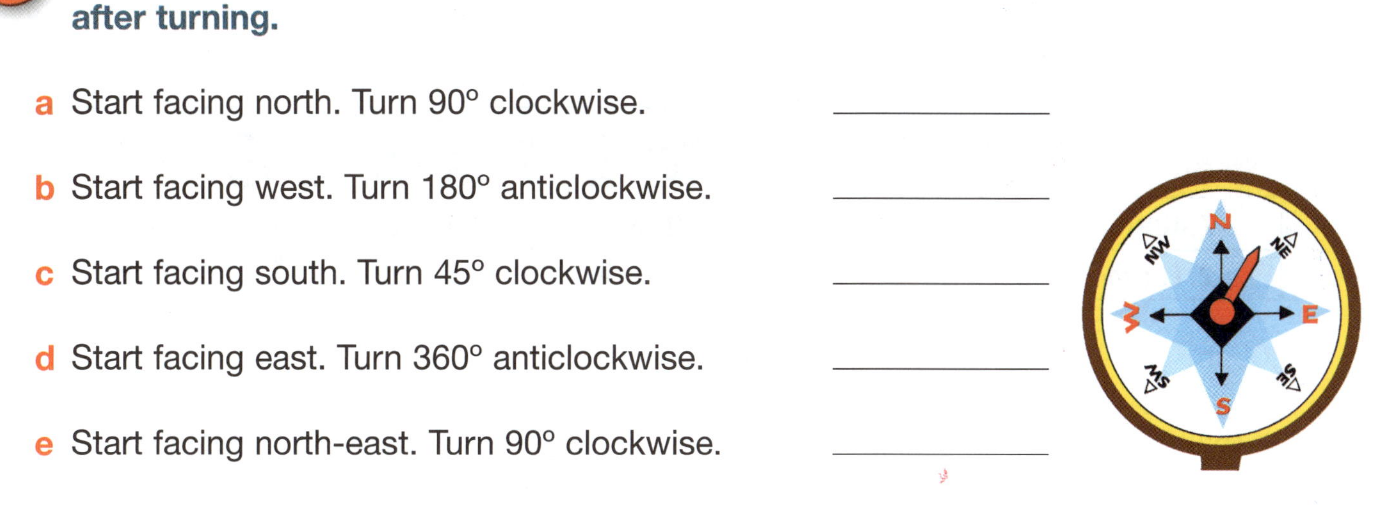

a Start facing north. Turn 90° clockwise. __________

b Start facing west. Turn 180° anticlockwise. __________

c Start facing south. Turn 45° clockwise. __________

d Start facing east. Turn 360° anticlockwise. __________

e Start facing north-east. Turn 90° clockwise. __________

f Start facing north-west. Turn 45° anti-clockwise. __________

Time problems

There are 60 minutes in an hour and 24 hours in a day.

am stands for **ante meridiem** and means **before midday.**

pm stands for **post meridiem** and means **after midday.**

5.35am

35 minutes past 5 in the morning

7.15pm

15 minutes past 7 in the evening

I **Draw the hands on the clock or write the digital time for each start and finish time.**

Start | **Finish**

a  Mark goes swimming at 10.15am. He gets home $1\frac{1}{2}$ hours later. 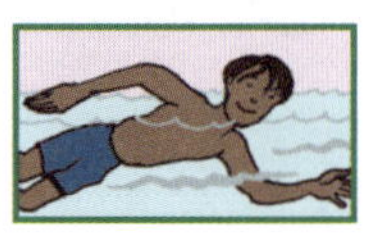 

b 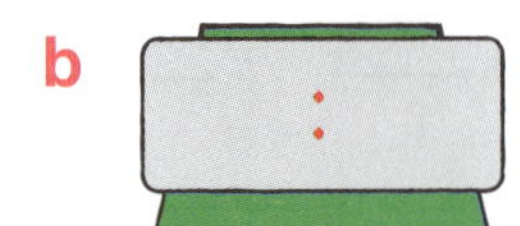 A train leaves London at 6.20pm. It arrives at Leeds 2 hours 20 minutes later. 

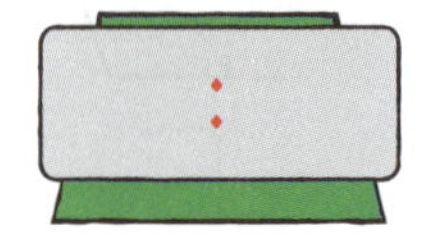

c 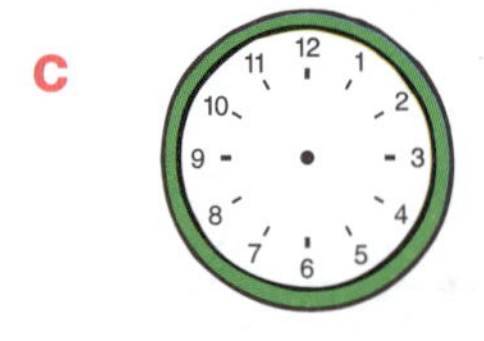 Becky goes shopping at 11.10am. She finishes 3 hours 45 minutes later.  

d 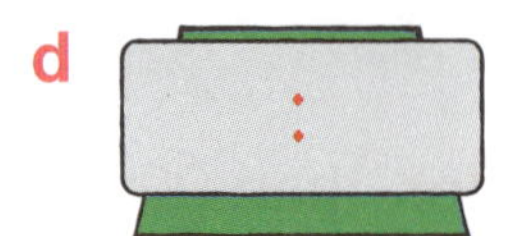 A football match starts at 1.45pm. It finishes 90 minutes later.  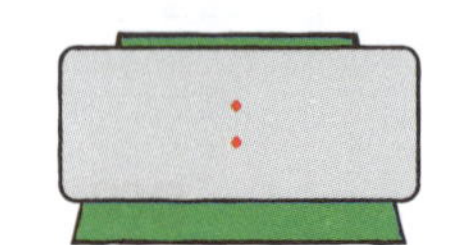

II **This timetable shows the times of buses. If you are at a bus stop at these times, how long will you have to wait?**

BUS TIMETABLE				
7.40am	8.15am	9.20am	10.50am	11.40am
2.10pm	4.30pm	5.10pm	6.30pm	8.00pm

a 9.05am = ______ minutes

b 11.15am = ______ minutes

c 5.05pm = ______ minutes

d 10.35am = ______ minutes

e 7.40pm = ______ minutes

f 4.45pm = ______ minutes

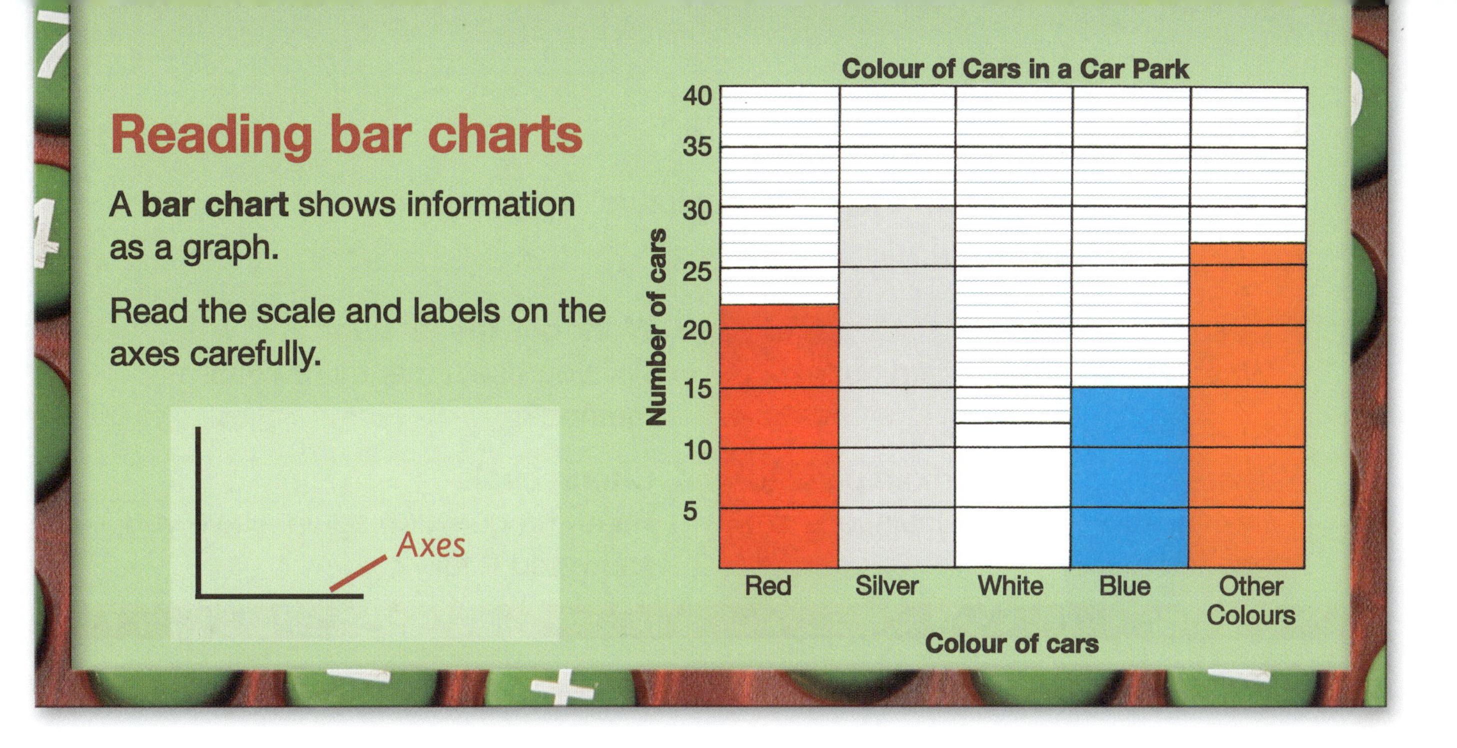

I Look at the graph above and answer these.

a Which colour was the most common car colour in the car park? __________

b How many cars were white? __________

c How many more cars were red than white? __________

d Which colour had half the number of silver cars? __________

e Black was the most common 'Other colour', with $\frac{1}{3}$ of these cars black. How many cars in total were black? __________

f How many cars in total were in the car park? __________

II This graph shows the number of cars visiting a car wash over five days.

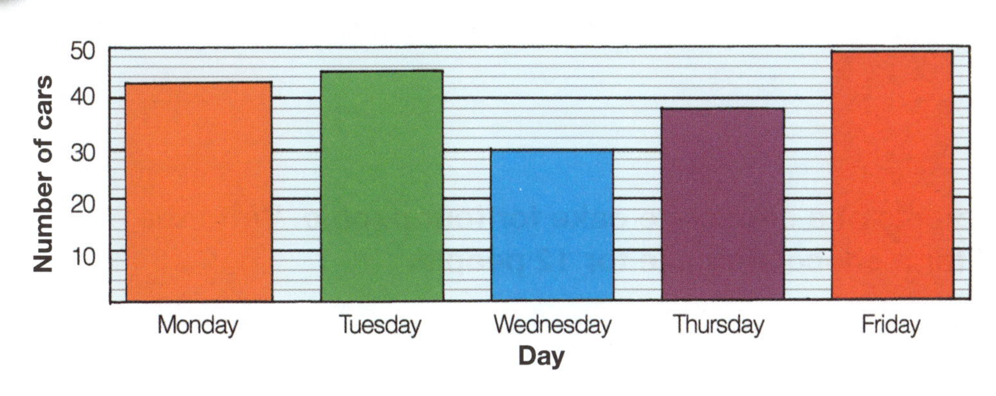

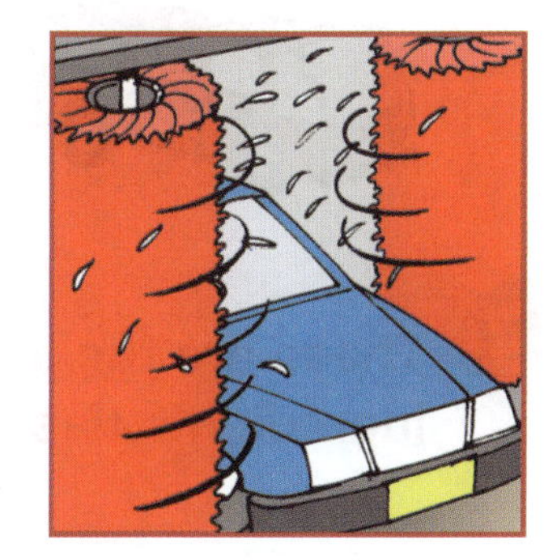

a How many cars visited the car wash on Tuesday? __________

b On which day did 38 cars visit the car wash? __________

c How many more cars visited on Friday than Monday? __________

d On which day did 15 fewer cars visit the car wash than on Tuesday? __________

Problems

When you read a **word problem**, try to 'picture' the problem.

Try these four steps.

1. **Read the problem**
   What do you need to find out?
2. **Sort out the calculation**
   There may be one or more parts to the question. What calculations are needed?
3. **Work out the answer**
   Will you use a mental or written method?
4. **Check back**
   Read the question again. Have you answered it fully?

I Read these word problems and answer them.

a A bar of chocolate costs 45p. What do 4 bars cost? ____________

b Sophie has 90 g of butter. She uses 35 g to make a loaf of bread. How much butter is left? ____________

c A board game costs £8.40. It is reduced by £2.50 in a sale. What is the new price of the game? ____________

d 68 people are going on a trip. Minibuses can take 10 people. How many minibuses will be needed? ____________

e A pencil costs 19p. How many can be bought for £2? ____________

f Sam has £39 to spend on cinema tickets. If cinema tickets cost £4, how many tickets can he buy? ____________

g Mrs Benson travels 48 km each day to get to work and back. How far will she travel in 5 days? ____________

h 73 sheets of paper are put into folders that each hold 5 sheets of paper. How many folders are needed? ____________

II These are the ingredients of a chocolate cake for four people. Write the ingredients needed for a chocolate cake for 12 people.

50 g margarine
40 g sugar
60 g flour
1 egg
15 g cocoa powder
20 ml milk

____
____
____
____
____
____

Coordinates

Coordinates help to **find a position** on a grid.

Look at the coordinates of A and B.

Read the numbers across **horizontally** and then up **vertically** for the pair of coordinates. (2,6) and (7,3)

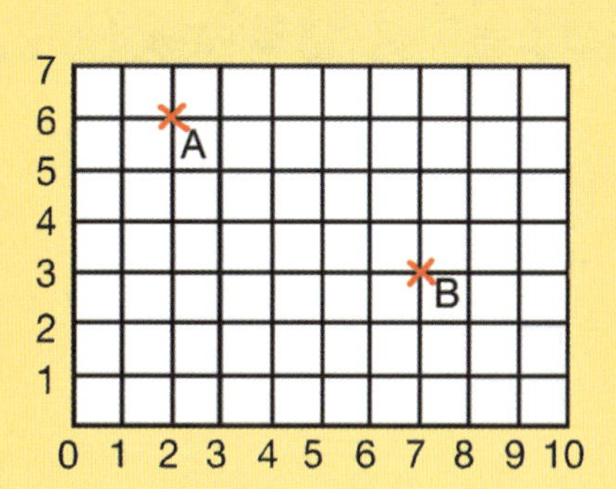

I Look at the grid below and answer the questions.

a What letter is at position:

(2,3) ___ (8,2) ___ (10,9) ___

b What are the coordinates for

D (___,___) A (___,___) S (___,___)

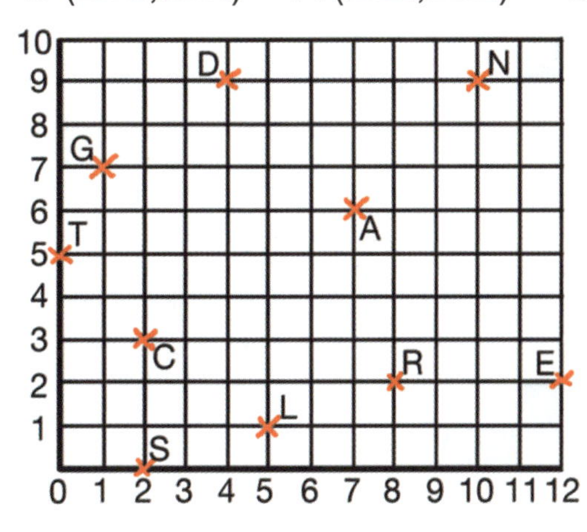

c Use the coordinates to spell out a shape and draw it in the box below.

(8,2) (12,2) (2,3) (0,5) (7,6) (10,9) (1,7) (5,1) (12,2)

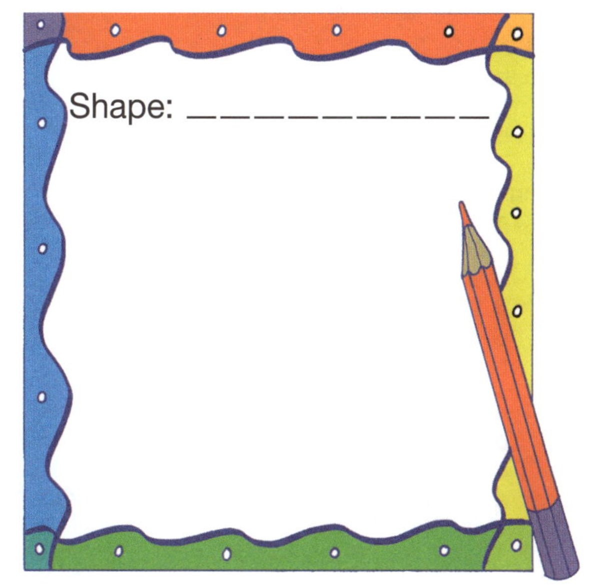

II Draw a quadrilateral on this grid.

The coordinates are:

(4,2) (6,5) (3,7) (1,4)

The shape is a ________________.

Move two of the coordinates to make the shape into a rectangle.

Write the coordinates of your rectangle.

__________ __________

__________ __________

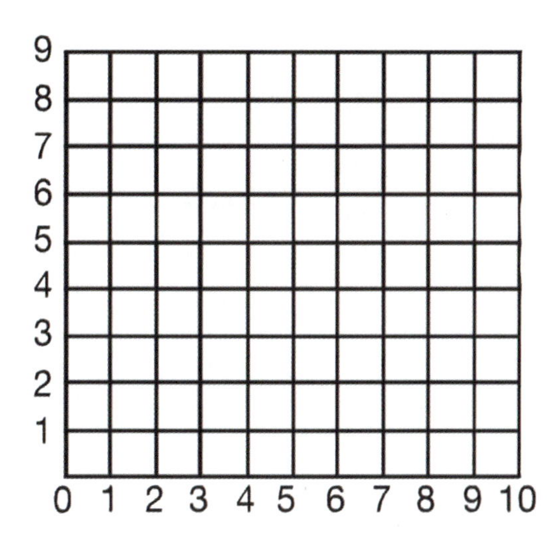

ANSWERS

Page 2

I a 758, 759, 760
b 618, 608, 598
c 506, 516, 526
d 741, 841, 941
e 285, 185, 85
f 902, 901, 900

II

	1 7	2 4	3	
3 3	■	0	■	4 6
5	■	5 9	2	0
1				8

Page 3

I a 29, 41, 45, 49 rule +4
b 300, 290, 280, 270 rule –10
c 144, 140, 136, 132 rule –2
d 77, 83, 86, 92 rule +3
e 112, 107, 102, 92 rule –5
f 40, 46, 58, 76 rule +6

II a –4, –3, –1, 1, 2
b –3, –2, –1, 2, 3
c –7, –6, –5, –3, –1, 0
d –2, –1, 0, 1, 2, 4

Page 4

I a 5 tens
b 6 thousands
c 8 ones
d 2 hundreds
e 6 tens
f 5 thousands
g 9 ones
h 7 hundreds
i 3 tens
j 4 ones
k 8 thousands
l 7 hundreds

II a 3850
b 7900
c 3680
d 4120
e 9000
f 465
g 291
h 340
i 507
j 800

Page 5

I a 12, 120, 1200
b 15, 150, 1500
c 15, 150, 1500
d 7, 70, 700
e 8, 80, 800
f 9, 90, 900
g 120
h 1300
i 1300
j 1800
k 240
l 40
m 1600

II

34	51	59	41	76	82	38	62
66	75	25	65	24	47	53	77
91	19	72	83	17	45	96	13
24	81	74	56	35	55	48	52

Page 6

I a triangle ✔
b octagon ✔
c pentagon
d octagon
e heptagon
f hexagon
g quadrilateral
h pentagon ✔
i hexagon ✔
j decagon
k nonagon
l triangle

II For a, b, c, d and h, there are many possible answers. Check each shape has the correct number of sides.
d check there is a right angle.
e
f
g

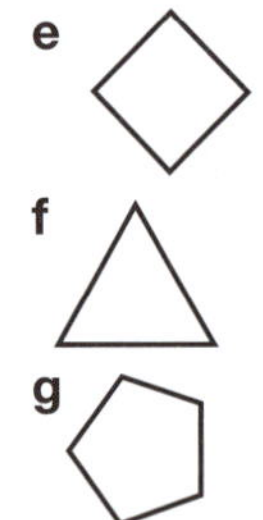

Page 7

I a £958, £1090, £1900, £2589, £2850
b 965 km, 2830 km, 3095 km, 3520 km, 3755 km
c 1599 g, 1995 g, 2046 g, 2460 g, 2604 g
d 4599 ml, 4600 ml, 7025 ml, 7028 ml, 7529 ml

II 2389, 2398, 2839, 2893, 2938, 2983
3289, 3298, 3829, 3892, 3928, 3982
8239, 8293, 8329, 8392, 8923, 8932
9238, 9283, 9328, 9382, 9823, 9832

Page 8

I a 4.12
b 10.38
c 12.26
d 5.34
e

f

g
h

II a 35 minutes
b 40 minutes
c 40 minutes
d 55 minutes

Page 9

I a 6, 3, 4, 2
b 10, 5, 4, 2
c 12, 8, 6, 4

II a any 18 squares red, any 9 blue, any 6 yellow. 3 left white.
b any 24 squares red, any 12 blue, any 8 yellow. 4 left white.

Page 10

I a 3500 m
b 4 cm
c 1.5 or $1\frac{1}{2}$ m
d 80 mm
e 25 cm
f 6.5 or $6\frac{1}{2}$ km
g 220 mm
h 1800 m
i 475 cm
j 6.5 or $6\frac{1}{2}$ cm

II a 45mm
b 62 mm
c 58 mm
d 37 mm
e 71 mm

Page 11

I a $3 \times 8 = 24$
$8 \times 3 = 24$
$24 \div 3 = 8$
$24 \div 8 = 3$
b $7 \times 4 = 28$
$4 \times 7 = 28$
$28 \div 4 = 7$
$28 \div 7 = 4$
c 6
d 4
e 21
f 5
g 6
h 24
i 4
j 8

II $60 \div 9 \rightarrow 6$
$93 \div 10 \rightarrow 3$
$89 \div 5 \rightarrow 4$
$38 \div 6 \rightarrow 2$
$106 \div 10 \rightarrow 6$
$61 \div 2 \rightarrow 1$
$37 \div 3 \rightarrow 1$
$48 \div 5 \rightarrow 3$
$65 \div 6 \rightarrow 5$
$80 \div 3 \rightarrow 2$
$53 \div 6 \rightarrow 5$
$46 \div 6 \rightarrow 4$

Page 12

I a > b < c >
d < e > f >
g > h < i <
j > k < l >

II a 4168, 4167, 4166, 4165
b 3839, 3840, 3841,
c 9001, 9000, 8999, 8998, 8997
d 4422, 4423, 4424, 4425
e 7081, 7080, 7079, 7078, 7077

Page 13

I a cuboid
b cylinder
c cone
d sphere
e cube
f pyramid

II

	faces	edges	vertices
a	5	8	5
b	6	12	8
c	5	9	6
d	4	6	4

Page 14

I a 2 kg
b 1500 g
c 5.5 or $5\frac{1}{2}$ kg
d 1.25 or $1\frac{1}{4}$ kg
e 7000 g
f 3250 g
g 10 000 g
h 6.75 or $6\frac{3}{4}$ kg
i 2500 g
j 4750 g
k 9.5 or $9\frac{1}{2}$ kg
l 1750 g

II a 2.5 or $2\frac{1}{2}$ kg
b 8 kg
c 6.5 or $6\frac{1}{2}$ kg
d 550 g
e 800 g
f 300 g

Page 15

I a 94
b 101
c 64
d 148
e 165
f 126
g 131
h 141
i 171
j 256
k 251
l 301